THE THEORY OF EVOLUTION IS INSANE

DANNY ERHARDT

The Theory of Evolution is Insane by Danny Erhardt

E-book: 979-8-3302-9710-8
Paperback: 979-8-3302-9709-2
Hardback: 979-8-3302-9711-5

Printed in the United States of America.

Genre Library Solutions
300 Delaware Ave. Suite 210,
Wilmington, DE 19801
www.genrelibrarysolutions.com
(315) 367-7314

CONTENTS

FORWARD

This book originally was written as a chapter (chapter 10) in a larger book about the truth titled "All the Secrets of the Universe (All the Ones worth Knowing): A Graduate Course on the Truth." And we have included the Introduction and Table of Contents of that book as an addendum at the end of this one. As a tease for those who might be interested in delving a little deeper into this search for the truth on other important subjects, after having their eyes opened on this one. That book, "All the Secrets of the Universe," systematically exposes all of the many lies of one called the great deceiver; lies that cover this earth like a blanket, with people basically believing whatever they are taught. (No matter how idiotic those beliefs may be: like the 72 virgin's Islamic scam; or the Theory of Evolution hoax.) And by exposing these very popular and widely held beliefs for the lies that they are, it affords one the opportunity to come to the knowledge of the truth. The truth, not based on opinion, speculation, faith, belief, religion, ideology, conjecture or theory, but based on all the facts and all the evidence, both scientifically and historically.

Here we have decided to offer this one chapter, number ten in the larger work, as a stand-alone book in and of itself due to the extraordinary importance of this information in a country, and a world, that has its collective head so firmly planted up its derriere when it comes to the bald-faced, and easily refuted lies surrounding this subject of Origins. Where did the universe come from? And how? How about life on earth? From where? And how? And the "current wisdom," the asinine but widely accepted explanation for these and other related questions, firmly held by the intellectual elites of our culture- in the media, the entertainment industry, and our educational system- is absurd. "The theory of evolution is *settled science*." (Yea, and my ass is the third reincarnation of Buddha.) Of course that oft-repeated "settled science" statement is one of the biggest whoppers ever perpetrated. Right up there with such historical ones like "YES WE CAN!" (Yes you can what? …fundamentally DESTROY America?? …Yes. Unfortunately. You can.) And "If you like your Doctor, you can keep your Doctor." And this total nonsense (and non-science) has been shouted from the mountaintops by these intellectual idiots…er…*elites* to the point where many have come to buy it. Like those who bought into Bernie Madoffs investment *opportunities.* As the Jew-killer Hitler and his henchman said, "If you shout a lie loud enough and long enough, enough stupid people will believe it." And in this regard, the big lie of evolution, we have all been equally snookered. And bamboozled. (… What say you, Fox News?? Come in Tucker? Hannity? O'Reilly? Goldberg?)

So here we offer, for your reading pleasure, Chapter 10 from All the Secrets. May it thoroughly wash your mind from the asinine and nearly insane deceit that is Satan's genesis, the Theory of Evolution....

INTRODUCTION

"In the beginning was the Word, and the Word was with God, and the Word was God. The same was in the beginning with God. **All things were made by him; and without him nothing was made that has been made."** (John 1:1-3)

Without Him, __nothing__ that you see in existence today was made. Nothing! Not by random accidental chance. Not by random accidental mutations. Not by a big explosion and random accidental collisions of matter in space. Nothing! It is not The Word of God that is *anti-science*—the Big Lie that the totalitarian-Democrat Left that controls the dissemination of information in our culture has been shouting for so long—it is the theory of evolution that is anti-science. In fact, as we will see in this chapter (book), it is a <u>scientific</u> **fact** that the theory of evolution is absolutely scientifically impossible according to the laws of science and all of the facts and evidence. (And remember, as Plato has instructed, we are following the facts and the evidence wherever they lead.) We will also see that it is so scientifically-impossible, so idiotic, so asinine, absurd and ridiculous, and such a flat-out, bald-faced lie, that it literally borders on the insane. (With all due respect to those who, like myself, were taught it and believed it.) The theory of evolution is nothing more than a false religion. It is "a fairy tale masquerading as science."[1] It is Satan's genesis, and it has everything to do with his plan to keep people from coming to the knowledge of the truth of Jesus Christ as their Lord and Savior by undermining the credibility of God's word by attacking the real origins account, the Bibles Genesis, and replacing it with his own idiotic lie.[2] And even though we have begun this chapter with reference to the <u>real</u> Creator, our conclusions actually have nothing to do with religion or theology. Religion and theology *are* indeed very important subjects that are indispensable to the discussion of the *real* origin of the universe, of life on earth, and the astronomical complexity and diversity of species. But

---

1   From a Creation Studies Institute newsletter, Sept 2013

2   (Evolution = Evil delusion)

neither of these subjects are necessary for exposing the outright un-scientific hoax—the world's greatest hoax—that is the theory of evolution. What the indisputable and overwhelming facts and evidence and information that we are going to present in this book have already clearly established beyond any reasonable doubt is that based on the immutable laws of science, and on all of the scientific facts and evidence, according to science itself and the scientific method alone, the theory of evolution is a joke.

And in addition, the same holds true for its companion theory, that of the big bang, which seeks to explain in purely natural terms the origin and "evolution" of the universe and our solar system from some mythical massive primordial explosion, out of which the entire universe just formed itself somehow, and which is equally as un-scientific and absurd. We will expose these twin hoaxes along with setting the record straight on a number of other related subjects such as the fossil record, Dinosaurs, anthropology, the worldwide flood, the speed of light and the ridiculous yet widely accepted notion that the universe and our solar system are billions of years old. These incomprehensible ages did not become our scientific and educational establishments preferred doctrine as a result of the unbiased search for scientific truth, but rather were the product of the desperate need that evolution had for virtually limitless periods of time in which to work its "miracle." In this case *necessity* truly was *the mother of invention.*

Macro vs. micro evolution

"As you know not what is the way of the spirit, nor how the bones do grow in the womb of her that is with child: even so you know not **the works of God, who made everything.**" (Ecclesiastes 11:5)

First it is important to understand the difference between these two terms, macro and micro evolution. Charles Darwin in his book, *On the Origin of Species by Means of Natural Selection, or the Preservation of Favoured Races in the Struggle for Life,* first published 150 years ago in November of 1859, attempted to show that the mind boggling diversity of life on earth happened by purely natural means. That by slow minor adaptations to their environment, each living species had slowly accumulated more advantageous characteristics and therefore had survived, whereas those competitors which had *not* "evolved," or somehow "self-created," those or other beneficial changes had become extinct. How *exactly* this could happen scientifically, the turning of one species into another entirely new, different and more complex organism by accident, was never explained by Darwin nor by anyone ever since; except to say with the devotion of pure, <u>blind</u> faith that "chance mutations over vast periods of time" did the trick. Nor did Darwin come up with an even remotely plausible scientific explanation as to *how* life could have accidentally formed itself originally, and nor has anyone since; except to say with the zeal of unwavering <u>blind</u> faith that somehow "lightning striking an ancient mythological atmosphere, or sea" did the trick. But the fundamental error of Darwin that led him to his erroneous conclusions in the first place was that he mistook *micro*evolution (<u>limited</u> change within one *kind* of organism), which is common to all of life, as being evidence for *macro*evolution (the massive change of one kind of organism into an entirely new and different one), which has never been observed, nor, as we shall see, has any credible evidence ever been found in the fossil record, or anywhere else on earth, to support it.

> "Micro-evolution [is defined as] small adaptations within a population of organisms which allow a certain trait to be expressed to a greater or lesser degree than before; variation within a given category. This is regularly observed to occur within living populations.
>
> "Macro-evolution [is defined as] large hypothetical changes which are thought to occur in an individual or in a population of organisms that produce an entirely new category or novel trait. These changes have never been observed to occur within living populations."[3]

…Or within dead populations (the fossil record) for that matter.

> "Often textbooks and newspapers cloud the truth by confusing terms. While sometimes referred to simply as "change," evolution …implies *enormous* changes in organisms. People supposedly came from ape-like ancestors, which came from a rat-like animal, which came from a fish, which came from an animal with no skeleton, which came from a single-celled organism. The proper term for this large-scale change is macroevolution. This *macro*evolution has… never been observed.

---

3   John Morris, Ph.D., *The Young Earth: The Real History of the Earth – Past, Present, and Future* (Green Forest, AR: Master Books, 2007), 8.

And keep in mind that the very *definition* of science is "the study of the physical and natural world and phenomena, especially by using systematic **observation**."[4] (Emphasis mine.) Macro-evolution has never been observed, therefore by the very *definition* of science, evolution is *not* science. It is a belief. And, as we shall see, like the Muslim suicide-bomber's faith that he'll get his 72 virgins, it is an idiotic belief that is not based in reality.

> "Frequently, what is observed scientifically is *micro*-evolution. Microevolution refers to small changes within an animal type. No new features are produced, but varying traits are expressed. This is also called adaptation or variation. *All* the proposed examples of observed change are *micro*evolutionary changes.
>
> "Scientifically proven examples of microevolution are: certain microbes' acquired resistance to antibiotics, pesticide-resistant insects, a color pattern shift in the peppered moth in England, and taller people today than those who lived long ago. Do these changes tell us anything about how people evolved from fish? Obviously not!"[5]

So microevolution, small changes within a kind, happen all the time. But this fact has absolutely nothing to do with, nor does it support in any way whatsoever, the scientifically-impossible theory that one organism can develop into another, higher, more advanced organism through accidental changes in its DNA. Macroevolution, as we shall prove throughout this chapter, is absolutely impossible. Never has occurred, never will. Back to the drawing board, Mr. Dawkins. We have big doggies and little doggies, short haired and long haired, Cocker Spaniels and Pit Bulls, Great Danes and Dalmatians. And we have tall humans and short humans, smart and dumb, black and white, Chinese and Indian, Eskimo and Pygmy. That's micro evolution; variation within a *kind*. ("And God said, Let the earth bring forth the living creature after its kind, cattle, and creeping thing, and beast of the earth each according to its kind: and it was so. And God made the beast of the earth after its kind, and cattle after their kind, and every thing that creeps upon the earth after its kind: and God saw that it was good." Genesis 1:24-25) But we don't have any flying humans. And there aren't any talking doggies. And that's because macroevolution doesn't exist in the real world, and

---

4   (Encarta Dictionary)

5   John D. Morris, Ph.D., *Dinosaurs, the Lost World, and You*, 6.

Darwin's theory of evolution, with all due respect to misinformed scientists everywhere, is impossible.

But what about mutations? Don't these cause major changes in an organism's DNA and couldn't they eventually change one living thing into an entirely new one? No, not a chance…

> "One possible source of significant change is that of mutations – the sudden, random, altering of a particular gene, which could then be passed on to offspring. Again, true science must ask, "What has been observed?" Although thousands of mutations have been observed, and thousands more have been induced, **never once has a mutation benefited an organism** or produced any *new* trait or functioning organ. **No mutation which could contribute to macroevolution is known to science.** Most mutations are neutral, many are harmful, and some are fatal. Therefore, it appears that random chance in the genetic code cannot improve it."[6] (Emphasis mine.)

"No mutation which could contribute to macroevolution is known to science."

Beyond that, according to the laws of mathematical probability, it is abundantly clear, as we shall see later, that "random chance in the genetic code" has no chance whatsoever of improving it, or of changing one viable organism into another. It is simply mathematically impossible for this to happen accidentally by random chance. Most of the systems in living organisms are irreducibly complex,[7] including most of the complex functions within the cell itself, and as such they could never have been produced by "small, incremental changes" (random mutations) within the genetic code. (See later section, "Irreducible Complexity.")

So, when we speak of *evolution* in this book, we are referring to <u>macro</u> evolution, which we shall show is scientifically impossible; and not to <u>micro</u> evolution (small alterations, adaptations, variations and changes within a

---

6   Morris, *Dinosaurs, the Lost World, and You*, 6-7.

7   (This is a term coined by Dr. Michael J. Behe, author of *Darwin's Black Box: The Biochemical Challenge to Evolution* and *The Edge of Evolution: The Search for the Limits of Darwinism*. We will discuss its fatal ramifications for the theory of evolution later in this chapter, in the section also titled, "Irreducible Complexity,".)

species or *kind*) which occur frequently but have never and could never lead to macroevolution.

Life is not the transforming miracle that evolutionists think it is. Like a much, much slower, organic version of the changing autobot robots in the movie *Transformers*, evolutionists suppose that life can just morph from one distinct life form into another, seemingly effortlessly, if given enough time. They make this gross misjudgment by observing the changes within a biological *kind* (read: species) that are pre-programmed into each organisms DNA, and surmising that therefore these organisms can miraculously transform themselves into whatever type of organism they might wish, again over vast periods of time. But, as we shall see, this is absolutely scientifically impossible. There is near infinite variety in living forms, and each of these living forms have pre-programmed into their DNA the ability for a fair degree of variation in response to their environments. (And this is because their Maker was not an idiot. He is a genius to dwarf all ingenuity, and He created his living things to be able to adapt and survive.) But there is great limitation within each organisms DNA as to how far that adaptive change is allowed to go. And it is <u>not</u> written in there to allow it to *transform* into another different *kind* of organism. Nor, as we just reported and as we will show in greater depth, can accidental mutations ever cause these supposed miraculous transformations. Again, life is <u>not</u> the transforming miracle that evolutionists suppose, and erroneously teach, that it is.

"And they shall turn away their ears from the truth, and shall be turned unto fables." (2 Timothy 4:4)

PART I

THE GREATEST HOAX ON EARTH

"And for this reason God shall send them a strong delusion, that they shall believe the lie." (2 Thessalonians 2:11)

"*The theory of evolution is a fact!*"[8] We have heard this stated so many times and so often by the high priests of this faux-science (a false religion disguised as science) in our educational system, and by their followers in our media and entertainment industry, that it has become ingrained in the psyche of far too many. But in reality the exact opposite is actually true.

> "The idea of evolution has come to be so firmly entrenched in our educational system that most people assume it is true. Scientific facts are [falsely and deceptively] placed within this interpretive scheme. End of discussion! Remember and repeat. Never mind the fact that no one has ever seen evolution take place, neither have the fossils documented evolutionary trends in the past, scientific law refutes the whole idea of evolution, and evolution is contrary to logic. Many people intuitively suspect evolution is not true, but still "believe" it anyway, because it is all they've been taught. "All educated people believe in evolution," they're told. Only ignorant, bigoted Christian fundamentalists still deny it."[9]

Yet the truth is that it is a total lack of proper education which bamboozles people into believing in evolution in the first place, and the real

---

8   (Sure, and so is Santa Claus and his 9 flying reindeer.)

9   Morris, *The Young Earth*, 6.

1

ignorance and bigotry lies in those evolutionists who refuse to even entertain the thought that there might be another side to their fairy tale…er…story. ("See no evil, hear no evil." …See no truth, hear no truth.) But let's start at the beginning. Let's take a look at all of the "evidence" that has been taught for decades as *proof* for the theory of evolution. Keep in mind, as we will clearly show, that <u>real</u>, <u>scientific</u> evidence is nonexistent.

Fabrications, exaggerations, hoaxes and lies…

… the "evidence" for evolution.

- <u>Miller-Urey's 1952 "amino acid in a bottle" experiment</u> is still reported in classrooms to this day as evidence that the essential building-blocks-of-life just formed themselves accidentally from lightning striking inorganic chemicals in earth's "early" atmosphere. But that is absurd. First of all, there are a number of problems with their experiment that render it completely useless for those who believe in the accidental origin of life. One is that the gases used in their experiments were, with what geochemists know today, nothing like the gases that we could expect in a hypothetical "early" atmosphere of earth. (Again you have to assume an astronomically large age for the earth which flies in the face of all of the actual scientific <u>evidence</u> that is supportive of a far younger earth and solar system. See "The Young Earth" section later in the book.) "The concentrations of methane and ammonia were carefully selected [by the experimenters] to <u>insure</u> the production of organic molecules." (Emphasis mine.) But in the real world, "a methane-ammonia reducing atmosphere would be fatal to life forms." Also, Miller and Urey used mild sparks to "simulate" lightning but that "is unrealistic." Because "Actual lightning would have destroyed any organics which may have been present."[10] Back to the drawing board.

But all of that, as damning as it is, is irrelevant because it is a fact that even if the test tube conditions of Miller-Urey's experiment were exactly the same as in earth's supposed early atmosphere and included real lightning, it is absolutely impossible for the next step in the accidental formation of life to ever have occurred. And that is for those Miller-Urey amino acids (the simplest and most primitive of organic molecules) to just accidentally form themselves into the complex protein chains essential for cellular life. To believe in that sheer impossibility is to take a leap of blind faith comparable to finding a substance spewed from a volcano that resembles asphalt and then concluding that the entire interstate highway system made itself accidentally

---

10   Scott M. Huse, *The Collapse of Evolution* (Baker, 1986), 114.

by random chance. It would be like growing rock crystals in a jar and then teaching your students that this was proof that all of mankind's precisely cut gems and exquisitely manufactured jewelry just made themselves accidentally by random chance. That of course would be idiotic. But in those two cases no one has a vested interest in denying the existence of man as creator of highways or jewelry. In the case of life, evolutionists not only have a vested interest in denying the existence of God as Creator of life, their entire careers depend on it. Unfortunately for them, though, instead of being a strong evidence for the spontaneous formation of the building blocks of life and therefore the beginning steps of evolution, these early experiments of "life in a bottle," so highly touted then and still incredibly taught to this day, are evidence for the exact opposite. If the smartest scientific minds cannot recreate under completely controlled conditions in a laboratory the accidental formation of the necessary organic compounds for a living cell, then how in the hell could the earth have done so unconsciously?

> "The probability of getting hit by lightning is about 1 in 600,000 (fortunately). The probability of winning a lottery grand prize with a single ticket is about 1 in 5.2 million (unfortunately). …the chance formation of even the simplest replicating protein molecule is 1 in 10^{450}."[11]

That is ten with 450 zeroes behind it! And remember that it has been scientifically established that anything with odds greater than 1 in 10^{66} could <u>never</u> happen, even if the universe lasted for an infinite number of years. End of evolution.

> "Thus, we find that it is mathematically impossible for even the most elementary form of life to have arisen by mere chance. Life is no accident. It is not even something which brilliant scientists can synthesize. The bewildering complexity of even the most basic organic molecules completely rules out the chance of life originating apart from super-intelligent design and planning."[12]

To think that the essential building blocks of life, amino acids, could have somehow formed themselves accidentally and *then* somehow could

---

11   Huse, *The Collapse of Evolution*, 65, 68.

12   Huse, *The Collapse of Evolution*, 69.

have come together completely unconsciously to link up and create a single protein molecule (much less the very large number that are necessary for the life of a cell), is beyond preposterous. And this is not speculation, theory or conjecture. It is established scientific fact. So much for the fable that life just formed itself accidentally because a couple of scientists in the 1950's made an amino acid in a bottle.

But for those who prefer a more thorough analysis of this, we would recommend reading pgs. 113-115 in Scott M. Huse's *The Collapse of Evolution*; and the article "The Miller-Urey experiment," by J. H. John Peet, BSc, MSc, PhD, CChem, FRSC, online @ truthinscience.org.

- <u>Recapitulation theory</u> Another phony proof of evolution that we were all taught comes from the fudged drawings of a German zoology professor, Ernst Haeckel. He was an avid promoter of Darwinism and in 1866 advanced a theory that claimed the developing human embryo passes through all of its "previous evolutionary stages"—e.g. a worm, a fish with alleged gill slits, a frog and then reptile[13]—and "showed" this progression in a series of drawings. It is called *recapitulation theory* which states that "ontogeny recapitulates phylogeny," or an organism's embryonic development (ontogeny) mirrors its supposed *evolutionary path* (phylogeny). But it has long been known that Haeckel made up his diagrams, and the actual pictures we now have of human embryo development show his bald-faced distortions for what they are. His theory was and is a lie, but incredibly that still hasn't kept our public schools from continuing to use it to indoctrinate our children in the greater lie. (Would it be fair to ask, What the hail is going on here, in what is supposedly a "science" curriculum?!#$)

> "Eventually, Ernest Haeckel admitted this fraud, but the deplorable aspect is that this theory is still taught in many universities, schools, and colleges throughout the world."[14]

> "Evolutionists have taught for over a century that as an embryo develops, it passes through stages that mimic an evolutionary sequence. In other words, in a few weeks an unborn human repeats stages that supposedly took millions of years for mankind. A well-known example of this ridiculous teaching is that embryos of mammals have "gill slits," because mammals supposedly evolved from fish.

---

13   Huse, *The Collapse of Evolution,* 113.

14   Huse, *The Collapse of Evolution,* 105.

(Yes, that's faulty logic.) Embryonic tissues that resemble "gill slits" have nothing to do with breathing; they are neither gills nor slits. Instead, those embryonic tissues develop into parts of the face, bones of the middle ear, and endocrine glands.

"Embryologists no longer consider the superficial similarities between a few embryos and the adult forms of simpler animals as evidence for evolution. Ernest Haeckel, by deliberately falsifying his drawings, originated and popularized this incorrect but widespread belief. Many modern textbooks continue to spread this false idea as evidence for evolution."[15]

"Valid science must have integrity, dependability, reliability, and be trustworthy. How can you come to true conclusions when experimental data is falsified?"[16]

Incredibly, evolutionists even tried to raise this hoax to the status of a *law* of science by calling it *The Biogenetic Law*.[17] However, this "law"—which merely attempts, fraudulently, to elevate the status of Haeckel's bogus theory—is scientifically absurd:

"The field of molecular genetics establishes the impossibility of the Biogenetic "Law." DNA is very specific and uniquely programmed for each type of organism. It simply does not recreate passing developmental stages of other organisms. It only produces after its own kind.

"Almost all scientists now reject the Biogenetic "Law." Only naïve or poorly informed evolutionists still cite this concept in defense of their theory for it has no valid scientific foundation whatsoever."[18]

Scratch one more "evidence" for evolution.

---

15 Walt Brown, Ph.D., *In The Beginning: Compelling Evidence for Creation and the Flood* (Phoenix: Center for Scientific Creation, 2008), 11.

16 Tom DeRosa, *Evidence for Creation: Intelligent Answers for Open Minds* (Powder Springs, GA: Creation Book Publishers, 2001), 14.

17 (Which is not the same as the Law of Biogenesis, an actual law of science, "attributed to Louis Pasteur, [that] states that life arises from pre-existing life, not from nonliving material." From Wikipedia, the free encyclopedia.)

18 Huse, *The Collapse of Evolution*, 113.

<u>- England's peppered moths</u> But what about the case of England's famous peppered moths? Aren't they "proof" of evolution, as we were all taught?

> "Peppered moths have always existed in light, intermediate, and dark-colored varieties. Before the advance of the industrial revolution, the tree trunks were light-colored and the light-colored moths were camouflaged; whereas, the dark-colored moths were easily spotted and eaten by birds. Consequently, the dark colored moths constituted a very minor proportion of the total population.
>
> "As the industrial revolution progressed, however, and pollution increased, the tree trunks became darker and within 45 years the situation was reversed. In the Manchester vicinity, for example, 95% of the moths were of the dark-colored variety.
>
> "But is this really evolution? Certainly not! This process did not produce anything new. It did not result in increased complexity and organization. The dark-colored moths always existed. ...absolutely no evolutionary change occurred in these moths... Despite these obvious facts, many textbooks and encyclopedias continue to cite the peppered moth as an example of evolutionary development."[19]

England's peppered moths are an example of *micro*evolution, change within a kind, and not of Darwinian evolution. Through natural selection the fittest moths did survive; first the light-colored, and then as the trees darkened, the dark-colored. But they did <u>not</u> evolve into a new organism with any new, different or more complex organs; because that is impossible.

<u>- "Vestigial" organs</u> In the past evolutionists have tried to use the case of organs, like the thyroid, pituitary glands and the appendix, whose function scientists had yet to uncover, as evidence for their theory:

Evolutionists have incorrectly stated and taught for years that vestigial organs are evidence for evolution because, according to them, they were a useful organ in ancient, mythological times past. But that is false. Now as we have come to understand much more of physiology we have found that these supposed "vestigial" organs are not only needed but many times absolutely

---

19   Huse, *The Collapse of Evolution,* 108.

essential to the life of the organism. Currently, organs that were once classified as "vestigial" and useless have all been shown to perform some necessary function for the organism.[20]

    <u>- The Galapagos Islands</u> But how about Darwin's famous Galapagos finches? Aren't they indisputable proof for the "fact" of evolution? No, they have just been taught that way. They are just one of thousands of examples we see across this planet of *variation within a species*, or <u>micro</u>evolution. Again, all organisms have pre-programmed within their DNA the ability to change and alter and adapt to the changing circumstances of their environments. This is because their Creator had foresight. But they do <u>not</u> have the ability to change into another species slowly over time by many different accidental mutations in their genetic code. That is scientifically impossible.

> "Christians look to Jerusalem; Muslims look to Mecca; but evolutionists look to the Galapagos as the spiritual center of their scientific faith."[21]

When Darwin visited the Galapagos Islands in the 1800's his observations "helped convince him that all living things evolved from a common ancestor." But science has come a long way in the last 150 years, and Darwin's fanciful but scientifically unsound conclusion has been thoroughly debunked. We now know that the "simple single cell" is an astronomically complex engineering marvel that could never, ever have happened accidentally by random chance. And we now know that the "endless forms most beautiful and most wonderful" that Darwin erroneously concluded "have been, and are being, evolved,"[22] are in fact displaying the built-in ability for great variation and adaptation that their Designer/Creator wrote into their DNA blueprints. It turns out that the Galapagos Islands are *not* the hallowed "laboratory of evolution" that devout Darwinists have long-claimed, but rather are a stunning "showcase for creation."[23]

> "The random chaos that is the fundamental idea of evolution is altogether at odds with the rationality

---

20   Huse, *The Collapse of Evolution*, 107.

21   (Doug Phillips, executive producer of *The Mysterious Islands* DVD. -See list of recommended books, and DVD's, at end of chapter)

22   Charles Darwin, *On the Origin of the Species*

23   (From "Impact," November 2009, a circular from Coral Ridge Ministries)

and design that we observe in creation. The mechanisms proposed for evolution do not square with the tenets of true science. Imagination, exaggeration, and in some cases, pure deception, are the hallmarks of the evolutionary movement.

"Evolution has been advertised and marketed to the public as the only credible and intellectual explanation of the origins of the universe and all life. Many have accepted it without question because it has been dressed up with false scientific integrity. There are those, usually at the universities, that have zealously dogmatized the world with the canons of evolutionary thinking. They have become the priests in their own religion and work with an evangelistic energy to spread the faith of evolution. Yet [it is all a] lie."[24]

- Other evidence There are a number of other things that evolutionists have cited as "proof" for their unscientific theory, and that, even though they have all been discredited, they still teach. Space precludes our listing all of them here but you can read chapter 6, "Commonly Cited "Proofs" of Evolution," from Scott Huse's exposé, *The Collapse of Evolution*. Here is his conclusion:

Evolutionists have consistently and relentlessly offered as their "proofs" of evolution numerous things such as the *Law* of Biogenic, the duck-billed platypus, comparative anatomy, the peppered moth, Miller-Urey's experiment, and Archaeopteryx. But they have falsely assumed these concepts as proof when in reality there is no evidence to back up their conclusions. Certainly not in the fossil record. Again, so much for their "proofs" of evolution.[25]

And speaking of "the fossil record," that just happens to be our next subject…

The fossil record

"Before the mountains were brought forth, or ever you had formed the earth and the world, even from everlasting to everlasting, you are God." (Psalms 90:2)

But what about the fossil record? Isn't this, finally, strong proof of the theory of evolution? Not hardly. We were all taught that this was such a powerful evidence that proved the theory of evolution was a *fact*. But it is not.

---

24   DeRosa, *Evidence for Creation*, 69-70.

25   Huse, *The Collapse of Evolution*, 117.

The problem is that after literally millions of evolutionary scientists and their devotees have eagerly, expectantly and exhaustively searched for trillions of hours and for well over a century and all over the globe, they have not found a single missing link in the fossil record. And that's because <u>there aren't any</u>. If evolution was real than for every single species we see there would have to be countless transitional forms that slowly, eventually led to each one of those species. For every bird that supposedly successfully developed or "evolved" an actual functioning set of wings on its own, out of thin air, accidentally,[26] there would have to be many many fossils of semi-birds with really bad, unusable wings that died off in the blind and unconscious struggle for survival, and in this case the *search* for flight. (How an organism searches or "struggles" for something they don't even know exists is beyond me.) The same would have to be true of every species. But there are none whatsoever. All the fossils that we find are fully functioning, fully developed organisms. Even Charles Darwin, the author of the theory, stated that the fossil record would be the final authority as to the validity of his proposal. But the scientific fact of the total lack of evidence in the fossil record is conveniently ignored, so Satan's genesis can still be shoved into the minds of our unsuspecting school children.

Creation as observed and written in the Bible (and remember kids, "Science is observation") is undeniably confirmed by the fossil record that shows 1) a total lack of transitional organisms, 2) the absence of change in species over time, 3) the nonexistence of any evolutionary predecessors and 4) the sudden explosion of highly developed and countless species that appear, SHAZAM!, just like that. Where for Pete sakes, is the slow incremental change over time that is the hallmark (i.e. lie) of evolution. It's just not there. Because evolution doesn't exist.[27]

> "All species appear fully developed, not partly developed. They show design. There are no examples of half developed feathers, eyes, skin, tubes (arteries, veins, intestines, etc.), or any of the vital organs (dozens in human alone). Tubes that are not 100% complete are a liability; so are partially developed organs and some body parts. For

---

26   (Think for a moment for how many centuries, even millenniums, man with all of his intelligence and design capabilities strived to learn how to fly but failed miserably at every attempt until finally the Wright brothers were successful. And then try to convince yourself that birds and insects and lizards could have each stumbled upon flight accidentally. It is truly idiotic.)

27   Huse, *The Collapse of Evolution*, 54.

example, if a leg of a reptile were to evolve into a wing of a bird, it would become a bad leg long before it became a good wing."[28]

Besides, common sense itself dictates that virtually <u>any</u> complex organ could never be mutated from another previous, less-complex organ because along the mythological way from <u>A</u> (less-complex organ) to <u>Z</u> (more-complex organ), every letter (every *step*) in between would be a lousy alteration of <u>A</u> and an incomplete version of <u>Z</u> and therefore due to the survival of the fittest these organisms would be killed or die almost immediately. So therefore, even if it were possible for a random mutation to even get the transition from a lizard leg to a bird wing started (which it is not) it would merely result in a crippled leg and the lizard would be eaten as it tried to slunk away from its predator. And this holds true with any supposed advancement on Darwin's mythological tree of death (yes it is death and not life). Whether it's fish to snakes, or deer to giraffe, or elephant to whale, or reptile to bird, or bush to beaver. End of evolution. The only possible way for a random mutation to turn a lizard into a bird is to do it all at once in a single generation, and this would be indistinguishable from a miracle. And the theory of evolution is not based on miracles. (Creation is.) It is based on small incremental accidental mutations that slowly advance the organism forward into another better, more highly developed, and better adapted organism, which is logically, factually and scientifically impossible. So much so that it surpasses the ridiculous and the irrational and borders on the insane. Again, as we mentioned in the beginning of All the Secrets (the larger book that this book/chapter 10 there/ came from), we live in an insane asylum, and the inmates have taken over. Literally. And I don't say that arrogantly or judgmentally because I was taught it and I believed it along with everyone else I went to school with. The lies of Satan cover our educational/left-wing-Democrat indoctrination system like a muddy layer of sediment, and we all believed the idiocy that we were taught.

> "At the most fundamental level, a big gap exists between forms of life whose cells have nuclei (eukaryotes, such as plants, animals, and fungi) and those that don't (prokaryotes such as bacteria and blue-green algae). Fossil links are also missing between large groupings of plants, between single-celled forms of life and invertebrates (animals without backbones), among insects, between

---

28   Brown, *In the Beginning*, 7.

invertebrates and vertebrates (animals with backbones), between fish and amphibians, between amphibians and reptiles, between reptiles and mammals, between reptiles and birds, between primates and other mammals, and between apes and other primates. In fact, *chains* are missing, not *links*. The fossil record has been studied so thoroughly that it is safe to conclude that these gaps are real; they will never be filled."[29]

It is a flat-out deception to teach the school children of America that the fossil record is evidence for evolution and for vast periods of time.

"The "evolutionary tree" has no trunk. In what evolutionists call the earliest part of the fossil record (generally the lowest sedimentary layers of Cambrian rock), life appears suddenly, full-blown, complex, diversified, and dispersed—worldwide. Evolution predicts that minor variations should slowly accumulate, eventually becoming major categories of organisms. Instead, the opposite is found. Almost all of today's plant and animal phyla—including flowering plants, vascular plants, and vertebrates—appear at the base of the fossil record. **In fact, many more phyla are found in the Cambrian then exist today.** [Emphasis mine.] [Oops, back to the drawing board. Next theory: <u>d</u>evolution.] Complex species, such as fish, worms, corals, trilobites, jellyfish, sponges, mollusks, and brachiopods appear suddenly, with no sign anywhere on earth of gradual development from simpler forms. Insects, a class comprising four-fifths of all known animal species (living and extinct), have no known evolutionary ancestors. The fossil record does not support evolution.

…"Frequently, fossils are not vertically sequenced in the assumed evolutionary order. For example, in Uzbekistan, 86 consecutive hoofprints of horses were found in rocks dating back to the dinosaurs. Hoofprints of some other animal are alongside 1,000 dinosaur footprints in Virginia. A leading authority on the Grand Canyon published photographs of horselike footprints visible in rocks that, according to the theory of evolution, predate hoofed animals by more than 100 million years. [Oops!]

---

29   Brown, *In The Beginning*, 11-12.

Dinosaur and humanlike footprints were found together in Turkmenistan and Arizona. Sometimes, land animals, flying animals, and marine animals are fossilized side-by-side in the same rock. Dinosaur, whale, elephant, horse, and other fossils, plus crude human tools, have reportedly been found in phosphate beds in South Carolina. Coal beds contain round, black lumps called *coal balls*, some of which contain flowering plants that allegedly evolved 100 million years after the coal bed was formed. In the Grand Canyon, in Venezuela, in Kashmir, and in Guyana, spores of ferns and pollen from flowering plants are found in Cambrian rocks—rock supposedly deposited before flowering plants evolved. Pollen has also been found in Precambrian rocks deposited before life allegedly evolved. ["Houston, we have a problem."]

"Petrified trees in Arizona's Petrified Forest National Park contain fossilized nests of bees and cocoons of wasps. The petrified forests are reputedly 220 million years old, while bees (and flowering plants, which bees require) supposedly evolved almost 100 million years later. Pollinating insects and fossil flies, with long, well-developed tubes for sucking nectar from flowers, are dated 25 million years before flowers are assumed to have evolved. Most evolutionists and textbooks systematically ignore discoveries which conflict with the evolutionary time scale."[30]

Wait a minute! Don't pass by that last statement too quickly. "Most evolutionists and textbooks **systematically ignore** discoveries which conflict with the evolutionary time scale." Wow! Is that *science*, to systematically ignore evidence, to ignore observations? <u>No</u>, actually, that is what is known as <u>lying by omission</u>, and not science. To "systematically ignore discoveries" is to be ignorant, at least the last time I checked. And yet it happens all the time, every day, all across this great land, in the name of *science*. In the name of *evolutionary* science (oxymoron).

- <u>Living Fossils</u> And then there are "living fossils" ...

---

30   Brown, *In The Beginning*, 12.

"Did you know that flamingos, sandpipers, penguins, cormorants, parrots, owls and many other creatures living today, including numerous types of mammals, reptiles, amphibians and arthropods are found in supposedly 65-plus-million-year-old rock layers, when dinosaurs and other "pre-historic" beasts once roamed the earth? So, why don't we ever see them displayed in our museums or depicted together in books and textbooks?

Even though many fossils of these kinds of extent (or living) fauna have been found in allegedly "ancient" rock layers, this information is purposely censored from public view because it doesn't fit the evolutionist's millions-of-years' timeline."[31]

Ignoring the evidence in order to protect ones coveted, yet false theory; what does that have to do with science?

- The "Geologic Column" The Darwinists tell us that all across this world there is a "geologic column" of different sedimentary layers or strata that show us the orderly march of time, and of evolution, over hundreds of millions and even billions of years. They call this _uniformitarianism._ Yet nothing could be further from the truth. All the evidence in the sedimentary strata across this world cries out for their cataclysmic and sudden deposition through the process of _liquefaction_ (water saturated with earthen material that were deposited under enormous pressure), and the subsequent _rapid_ burial of an entire planets life forms, both animal and plant, from both land and sea. (See "A World Wide Catastrophic Flood" coming up later.)

The geologic column is based on their assuming that their theory of evolution is true, then they use that "column" as "proof" that their theory is true. This of course is circular reasoning. The order that they place the sedimentary formations in is according to their "index fossils." But their only reason for doing this is their belief that the theory of evolution is true. This is a joke because therefore, as Scott Huse says, "The primary evidence for evolution is the assumption of evolution!"[32] Insane! And they get away with shoving this UNscientific slop down the throats of our unsuspecting, gullible and naïve students in our colleges, universities, elementary and high

---

31 (From "Living Fossils" article in a Creation Studies Institute circular, June 2011, by Tom DeRosa, Executive Director and Founder)

32 Huse, _The Collapse of Evolution,_ 14-15.

schools. BTW, nowhere on this planet does their mythological *"geologic column"* actually exist. It's only in their minds. Incredible, isn't it? And they get away with calling it science.

> "Practically nowhere on earth can one find the so-called "geologic column." Most "geologic periods" are missing at most continental locations. Only 15-20% of the Earth's land surface has even one-third of these periods in the correct order. Even within the Grand Canyon, 150 million years of this imaginary column are missing. Using the assumed geologic column to date fossils and rocks is fallacious."[33] (...and stupid, and unscientific, and ignorant, and bigoted, and...)

There is no serious amount of fossilization happening across this earth today. Where are all the fossils of the buffaloes that we killed by the millions?? There are none. There is also no example whatsoever of the billions and trillions of marine life that die every day and every year becoming fossilized. They just rot, and/or are eaten. No fossil. This is a complete slap in the face (and in the brain) to uniformitarian evolutionists. Because we see a fossil record all across the planet, but obviously they could not have been formed according to the processes we see today. No uniformitarianism. Today any fossilization is rare, and not the norm. Therefore there has to be another reason for all of these fossils. And there is. They had to be formed by a worldwide rapid and very deep burial so they could be fossilized inside of the hardening sedimentary rock before they could be eaten or decompose. Hey kids, maybe the flood recorded in Genesis that occurred 4500 years ago, is true!! Gol-uhol-lee! Ya think?

That flood would also have sorted and laid down the organisms, and therefore their resulting fossils, from small to larger which of course has nothing to do with "evolutionary progression" but everything to do with creation and the flood.[34]

What would also be expected, and is also found, is countless anomalies in this fossil record where fossils of animals or advanced life forms are found in sedimentary strata hundreds of millions of years before they supposedly appeared/evolved. Who, or what, put them there? (Trump?) (And Russia?)

---

33   Brown, *In the Beginning,* 67.

34   Huse, *The Collapse of Evolution,* 47.

In addition to countless other examples of *polystratic*[35] fossilized trees, and even animals, going up through "millions of years" of sediment. Who put them there if it took millions of years for these sediments to be laid down? (You're right, Trump.)

"Geologists also have discovered polystrate animal fossils. Probably the most famous is the fossilized skeleton of a whale discovered in 1976 near Lompoc, California. The whale is covered in diatomaceous earth. Diatoms are microscopic algae. As diatoms die, their skeletons form deposits—a process that evolutionists say is extremely slow. But the whale (which is more than 75 feet long) is standing almost on its tail at an angle, and is completely covered by the diatomaceous earth. There is simply no way a whale could have stood upright for millions of years while diatoms covered it, because it would have decayed or been eaten by scavengers. It is clear from this extraordinary evidence that the long ages attached to the geologic column simply are not correct."[36]

"Fossils all over the world show evidence of rapid burial. Many fossils, such as fossilized jellyfish, show by the details of their soft, fleshy portions that they were buried rapidly, before they could decay. (Normally, dead animals and plants quickly decompose.) The presence of fossilized remains of many other adult animals, buried in mass graves and lying in twisted and contorted positions, suggests violent and rapid burial over large areas. These observations, together with the occurrence of compressed fossils and fossils that cut across two or more layers of sedimentary rock, are strong evidence that the sediments encasing these fossils were deposited rapidly—not over hundreds of millions of years. Furthermore, almost all sediments that form today's rocks were sorted by water. The worldwide fossil record is, therefore, evidence of rapid death and burial of animal and plant life by a worldwide,

---

35   (A *polystratic*, "many layers," fossil is a fossil that extends through many layers of strata, and therefore through "many eons of geologic time," which of course is impossible; but yet there they are, all over the world.)

36   Kyle Butt and Eric Lyons, *Dinosaurs Unleashed: The True Story about Dinosaurs and Humans* (Montgomery, AL: Apologetics Press, 2004), 55.

catastrophic flood. The fossil record is not evidence of slow change."[37]

The fossil record is full of animals and fish that died quickly and were buried rapidly. There is even a fossil of a fish *in the process* of swallowing another fish. Half of it is still sticking out of its mouth for goodness sake! How quickly and how deeply, in order to keep from being scavenged, did they have to be buried for that to happen? There are countless examples of fossils of animals and fish with their backs bowed downward because something very heavy came upon them suddenly, and buried them instantly. Also, if these sediments, and these fossils, were laid down over hundreds of millions of years then "the fossil record should show continuous and gradual changes from the bottom to the top layers." Yet there are vast gaps and many discontinuities throughout. The fossil record was supposed to be the Darwinists best friend, but truly it is his worst nightmare.[38]

> "Typically, fossils are found in fossil graveyards where the animals have been dismembered and the remains broken and scattered about. The bones are mixed with bones of other individual animals, making reconstruction extremely difficult. Animal and plant remains from vastly different environments and habitats are often mixed together. For instance, the dominant fossil at Dinosaur National Monument Park is a fossil clam. Many times, the deposits in which the fossils are found cover great areas on the earth's surface – at times hundreds of thousands of square miles. The event which deposited these sediments was not only of large geographic extent, but involved intense energy levels. Truly, what could have been larger and more powerful than the worldwide flood? [Hmm… evolutionists attempts to deny it?]

> "For any creature to be fossilized in a flood, it must be buried rapidly. Otherwise, it will decompose or be eaten within a fairly short period of time. Obviously, these vast fossil graveyards were collections of creatures who died in the global flood of Noah's day."[39]

---

37 Brown, *In the Beginning*, 11.

38 Brown, *In the Beginning*, 10-11.

39 Morris, *Dinosaurs, the Lost World, & You*, 34-35.

> "Fossils rarely form today, because dead plants and animals decay before they are buried in enough sediments to preserve their shapes. We certainly do not observe fossils forming in layered strata that can be traced over thousands of square miles. …[Only a worldwide deluge can explain] why animals and plants were trapped and buried in sediments that were quickly cemented to form the fossil record and why fossils of sea life are found on every major mountain range."[40]

It is also instructive that before the spread of the error of Darwinism, scientists all across this globe and for centuries <u>observed</u> the features of the earth and concluded that they were the result of a worldwide deluge. They could think of <u>nothing</u> else to explain them that made any sense whatsoever. And that's <u>not</u> because they were stupid. It's because they were not brainwashed and deceived by having it drilled into them that "the theory of evolution is a fact!" and "if you don't agree with us then you're stupid!" … like our professors today are. They did not *see* everything in the world through the distorted lenses of evolutionary glasses. They were not blind.

"Let them alone. They are blind leaders of the blind. And if the blind lead the blind, both shall fall into a ditch." (Matthew 15:14)

And for a further discussion on the actual cause of the world's fossils and sedimentary strata, see the section "A World Wide Cataclysmic Flood" later in this book.

Natural Selection

"You have made heaven, the heaven of heavens, with all their host, the earth and everything therein, the seas and all that is therein, and you preserve them all." (Nehemiah 9:6)

But what about survival of the fittest? Doesn't the struggle of all living things to survive cause evolution to occur? No, not hardly. It can cause only <u>micro</u>evolution to occur, where the ability to adapt to certain changing environmental pressures is already pre-programmed into the organisms DNA. And if that specific adaptive ability is not already preprogrammed then nothing, and no known mechanism, can cause it to miraculously appear within that organisms DNA. There is no question that in the wild the "strong do survive" and the weak perish, but this fact has nothing whatsoever to

---

40   Brown, *In the Beginning,* 116.

do with the idea that these *strong* can also develop, accidentally by random chance, more sophisticated and more highly advanced appendages or organs even to the point of morphing by accident into entirely new creatures that would allow them to be even *stronger*. For anything like that to ever happen, accidentally, within an organisms DNA is completely beyond the realm of possibility, or mathematical probability. It would be no different than if an old computer were able to change itself from Windows 95 to Windows 7 without a download or an upgrade to its software and hardware. Impossible. Again, were it to occur in either a computer or in life's DNA, it would be indistinguishable from a miracle.

> "Acquired characteristics—characteristics gained after birth cannot be inherited. For example, large muscles acquired by a man in a weight-lifting program cannot be inherited by his child. Nor did giraffes get long necks because their ancestors stretched to reach high leaves.

> ..."However, stressful environments for some animals and plants cause their offspring to express various defenses. New genetic traits are not created; instead, the environment can switch on **genetic machinery *already present*.** [Emphasis mine] The marvel is that optimal genetic machinery already exists to handle some contingencies, not that time, the environment, or "a need" can produce the machinery."[41]

> "Natural selection cannot produce *new* genes; it *selects* only among pre-existing characteristics. As the word "selection" implies, variations are reduced, not increased.

> ..."The variations Darwin observed among finches on different Galapagos islands is another example of natural selection producing micro- (**not** macro-) evolution. While natural selection sometimes explains the survival of the fittest, it does not explain the *origin* of the fittest. Today, some people think that because natural selection occurs, evolution must be correct. Actually, natural selection *prevents* major evolutionary changes."[42]

---

41   Brown, *In The Beginning*, 5-6.

42   Brown, *In The Beginning*, 6-7.

> "Mutations are the only known means by which new genetic material becomes available for evolution. Rarely, if ever, is a mutation beneficial to an organism in its natural environment. Almost all observable mutations are harmful; some are meaningless; many are lethal. No known mutation has ever produced a form of life having greater complexity and viability than its ancestors.
>
> …"A century of fruit fly experiments, and involving 3,000 consecutive generations, gives absolutely no basis for believing that any natural or artificial process can cause an increase in complexity and viability. No clear genetic improvement has ever been observed in any form of life, despite the many unnatural efforts to increase mutation rates."[43]

In addition, even if macroevolution were feasible (which it is not) and genetic mutations could somehow prove positive and then slowly change one organism into another over time with incremental alterations (which it also cannot), evolution would still be absolutely impossible for purely practical reasons. For (as we discussed previously, but let us reiterate) long before an organism changed *completely* into a more advanced and stronger form, it would have to try to survive as a transitional form which would have much <u>less</u> of an ability to survive than either the organism it was "struggling" away from, or the organism it was supposedly somehow blindly and unconsciously "struggling" towards. Let's look at the "classic" evolutionary example that reptiles, in particularly Archaeopteryx, evolved into birds…

> "How could a walking, ground-dwelling – or even tree-dwelling – reptile become a bird that could fly? [Why the same way that your pick-up truck could slowly, accidentally, become a Piper Cub; in the Alice-in-Wonderland fantasy world of the human mind.] For the front limbs to be effective wings, multiple mutations would have to result in progressively longer fingers or arms with a membrane between each finger. Long before the walking or climbing creature would have good wings, it would have inadequate legs and would be at a tremendous disadvantage against predators. In fact, it is hard to imagine

---

43   Brown, *In The Beginning*, 7.

how a creature that could not yet fly, but which possessed long fingers or arms, could even survive, let alone have an advantage in the struggle for existence. [Unless Darwinists were to come up with a variation of their theory, called "survival of the silliest." In this scenario predators would be laughing too hard to catch these prey, and the deformed transitions would therefore survive to "struggle" on...] Natural selection would weed out these mutant misfits.

[And that is why evolutionists, after scouring the globe for a hundred and fifty years, have not found one single transitional life form in the fossil remains—even though there should be trillions of them—not only because none ever existed, but because they could never have survived even _if_ they had.]

> "Furthermore the breathing apparatuses of birds and reptiles are entirely different. In birds, air flows in one air pipe and out another, while in reptiles, the air goes in and out through the same opening, as in mammals. Since breathing is essential for survival, and cannot be halted for even a few minutes, how could such a major transition occur by mutation and natural selection? [Answer: it could not.] A similar problem exists with the totally different design of the heart. Clearly birds did not evolve from dinosaurs or other reptiles."[44]

So, any way you slice it, and as kind and as bend-over-backwards accommodating as you might care to be, evolution still comes out smelling like dead fish. It's a joke.

The father and hero of modern evolutionists, Darwin, viewed "all beings, not as special creations, but as lineal descendants of some very few beings." And whereas that's a nice fantasy, it has no foundation in science, or in reality...

> "The complete title of his book, _The Origin of Species by Means of Natural Selection or the Preservation of Favored Races in the Struggle for Life_, illustrates the power that Darwin attributed to the concept of natural selection. His

---

44   Morris, _Dinosaurs, the Lost World, & You_, 22.

hope, centered on limited observations, was that the struggle for life would give us an upward movement of change in which the best of the best would succeed. ... [However] Darwin knew nothing of genetics and DNA in his day. He was forced to make suppositions about mechanisms at a much more simplistic level, but our understanding of genetics today actually invalidates those assumptions.

"The bedrock of Darwin's theory, "natural selection," is shown ...to have no creative power whatsoever. Natural selection only has the ability to eliminate populations with traits less suited to survival, or to cause traits that were already designed into the DNA to become more or less dominant, but it cannot create new characteristics (such as new body parts). Darwin's belief in the ability of organisms to acquire new characteristics through environmental influence, effort, exercise, use or disuse is also shown to be without merit. We know today that none of these factors has any influence upon the fixed coding of the DNA that is passed on to future offspring."[45]

Which is why a very wise individual said a very long time ago, "The fool has said in his heart, there is no God." (Psalms 14:1) Which translates into today's language as, "The improperly educated evolutionist has foolishly determined in his mind, there is no intelligent Creator." (Sciences 15:2)

On his visit to the Galapagos Islands in 1835, Charles Darwin observed...

..."changes and diversification. [But then] Darwin takes a great leap of faith: "From so simple a beginning endless forms most beautiful and most wonderful have been and are being evolved." Darwin concludes that small changes within a kind, something we can observe, are proof of something which has never been observed, evolution, or the transformation of one kind of animal into another."[46]

---

45   (From, Creation Studies Institute circular, October 2009, by Tom DeRosa)

46   (From, *The Mysterious Islands: A Surprising Journey to Darwin's Eden,* An Erwin Brothers Film)

However, we know a lot more now than the folks in Darwin's day, and it is clear from all the facts and evidence that we now have at our disposal that the natural selection of certain dominant or recessive traits within an organism, depending on the demands of its environment, is <u>not</u> proof of evolution. But it <u>is</u> proof that a Supreme intelligence was behind all of creation, and that the amazing power of *change, adaptation and variation <u>within</u> each species or biblical "kind"* is also proof that a supremely intelligent Creator was there pre-programming that adaptability into every species DNA. Just like the operating system in your computer is pre-programmed to handle a whole host of contingencies, and viruses, and errors.

Anthropology

"Then God said, Let us make man in our image, according to our likeness." (Genesis 1:26)

But what of anthropology? What of all of those charts that we see everywhere we go—in our schools, our museums, our universities and on our science channels—that show an ape slowly morphing into a man? And what of all those "ape-men" discoveries over the years, those "transitional" forms in between apes and men that have "proven" that man descended from apes? Answer: "Sorry, Wally World's closed." As far as the first question, the charts are fabrications. They are not based on science or the scientific evidence, but on the fertile imaginations and evolutionary biases of those who desperately want the unscientific idiocy of Darwinism to be true; and will tolerate no alternative. They have staked their entire lives and their reputations and $income$ on it, and what we see in their "ape-men" progression charts is their unscientific fantasies.

As far as the second question, there has not been found one single, solitary transitional form between apes and men. Every single *heralded discovery* over the years that claimed a fossil of a primitive and transitional ape-man was found, has turned out to be a gross exaggeration; either an outright hoax, or a chimpanzee, or an ape, or a pig. But our evolutionist friends who control the dissemination of information in our Democrat-Left media, educational system and entertainment industry have not thought it necessary to pass along this information to the public, or to their students. Like the truth about the Bible, or the Russian collusion hoax (spewed by the lying Democrats in our Congress and our media for 2 solid years), or the truth about an invasion across our southern border (labeled a "manufactured crisis" by those same lie-spewing, baby-mutilating, false-righteous, racist, black baby exterminating, nauseatingly-immoral Democrats)[47], this is all the news that is <u>not</u> deemed fit to print, or to share in any way with the unwashed masses. After all, they might come to the "wrong" conclusion… like evolution is a scientifically-impossible joke. Let's take a peek at these "transitional" forms, shall we?

"All flesh is not the same flesh, but there is one kind of flesh of men, another flesh of beasts, another of fishes, and another of birds." (1 Corinthians 15:39)

- <u>Piltdown "man"</u> In 1912 a few fragments of teeth, bones, and primitive implements were allegedly found in a gravel pit in Piltdown, Sussex, England. Anthropologists said the remains were over five hundred thousand years old. Countless piles of literature were written about Mr. Piltdown. It was celebrated in textbooks and in museums as an incredible discovery, as well as roughly five hundred doctoral dissertations being done on our "leetul fren" (Scarface).[48] But it was all a hoax…

> "[However] it is now universally acknowledged that Piltdown "man" was a hoax, yet Piltdown "man" was in textbooks for more than 40 years."[49]

Reader's Digest, in October, 1956, published an article, *The Great Piltdown Hoax* in which it reported the findings that the jaw-bone was that of an ape that had died 50 years before, and that the teeth had been filed down and biochromate of potash had been put on them in order to hide their identity. What a total deception!! But yet heralded by all of the "experts" as real, and with the utmost certainty. The culprit who fabricated the hoax wanted to synchronize Christianity with evolution by helping it along by creating a "missing link."[50]

How incredible is all of that? "<u>Over 500 doctoral dissertations</u> were performed on" a flat-out hoax!? "Piltdown man was viewed in stately museums and studied in major textbooks for several generations." Yet they were all viewing and studying a bald-faced lie! Darwinists were overcome with

---

47   (See the larger work, "All the Secrets of the Universe," for the truth about politically ideologies.)

48   Huse, *The Collapse of Evolution,* 100.

49   Brown, *In the Beginning,* 12.

50   Huse, *The Collapse of Evolution,* 100-101.

joy over their *great find*. It was such a wonderful *proof* of their hallowed theory, yet not one of them thought to actually <u>check it out</u> to see if it was valid, as actual scientists are supposed to do. No, they were too busy calling anyone who disagreed with them "stupid," "<u>un</u>scientific" and (worse yet) "religious." How pathetic. And how utterly opposed to the cause of true science.

One *transition* between apes and men down, just a few more to go…

<u>- Rama Lama Ding Dong</u> And then there's our old friend *Ramapithecus*, made famous by the Leakey's archeological work in Kenya, and splashed all over our TVs and newspapers for years as the missing link; and it was splashed confidently and without question by the evolutionary true believers who saturate our media. But, oops, it was nothing more than an ape:

> "Before 1977, evidence for *Ramapithecus* was a mere handful of teeth and jaw fragments. We now know these fragments were pieced together incorrectly by Louis Leakey and others into a form resembling part of the human jaw. … Some textbooks still claim that *Ramapithecus* is man's ancestor, an intermediate between man and some apelike ancestor. This mistaken belief resulted from piecing together, in 1932, fragments of upper teeth and bones into… two large pieces. …This was done so the shape of the jaw resembled the parabolic arch of man… In 1977, a complete lower jaw of *Ramapithecus* was found. The true shape of the jaw was not parabolic, but rather U-shaped, distinctive of apes. … *Ramapithecus* was just an ape."[51]

However, the problem here is that this is the truth. And Biology textbooks on evolution are exempt from the normal requirements of educational material, like printing the truth for example. It is a religion after all.

> "Fossils of supposedly ape-like man are speculative, incomplete, falsified and imagined."[52]

<u>- Nebraska "man"</u> Next the Cornhuskers have an entry:

Harold Cook was the discoverer of the famous Nebraska man in some "Pliocene" sediments there. It was said that this "missing link" was a million

---

51   Brown, *In the Beginning*, 13.

52   DeRosa, *Evidence for Creation*, 61.

years old and copious amounts of writings were published about it. [But, shocker, it was all a lie.]

Remember the famous Scopes trial in 1925 held in Dayton, Tennessee? Where evolutionists lambasted William Jennings Bryan with all of the "great scientific experts" and their "proofs" concerning that Nebraska man. That caught Bryan off guard because he didn't know enough at that time to realize that all of those "great scientific experts" were full of great amounts of s**t and so he asked for more time to respond. Of course the great "experts" ("liars") mocked and ridiculed him. "After all, who was he to question the world's greatest scientific authorities?" [Like the "great scientific authorities" who wrote those 500 plus doctoral dissertations on the Piltdown hoax.]

The really hilarious thing which unfortunately Mr. William Jennings Bryan did not know at the time was that the entire hoax that they used to laugh at him was built on a single tooth!! That was the great "find" that led to the "missing link" called Nebraska man!! The world's top evolutionists scrutinized that tooth and declared that it was absolute proof of a primitive species of man. What a classic case of a bunch of pompous, ignorant horse's asses hurling false accusations at a great American patriot, and slandering his character and reputation in the process. (Kind of reminds you of what the disgusting congressional Democrats and their lying Demonrat media tried to do to Trump, and Kavanaugh.)]

Some years later they found the entire skeleton that the tooth came from, and guess what?? It was an extinct pig!! Bryan was mocked for his "ignorance" by experts (ignoramuses) who fabricated a complete species of man from a pig's tooth!![53]

But, again, that's the truth. Which won't in any way hinder our public school system from continuing to indoctrinate our children (mostly out of ignorance I am sure, and hopefully not out of any intentional desire to deceive) in the falsehood that the Scopes trial was a "victory" for evolution; rather than teaching them the truth that it was an historical showcase for how those who are obsessed with a scientifically-impossible theory can make complete fools out of themselves by combining just one pigs tooth with their over-heated imagination, and coming up with an entirely new and non-existent species in the process.

- A cup of Java Another false alarm was *Pithecanthropus erectus* (not to be confused with an erectile dysfunction drug of a similar name) or Java

---

53   Huse, *The Collapse of Evolution*, 97-98.

Ape-Man. In fact, it was the discoverer, Eugene Dubois himself who forty years later…

> …"conceded that it was not a man, but was similar to a large gibbon (an ape). In citing evidence to support this new conclusion, Dubois admitted that he had withheld parts of four other thigh bones of apes found in the same area."[54]

The suspicious pronouncements surrounding Java Ape-Man are hidden behind the façade (the lie) of "facts" of science. Indeed, in 1926 an additional *Pithecanthropus erectus* find was also heralded as THE missing link (finally!!), and such a "wonderful discovery"! But, OOPS, it was nothing more than a now-extinct elephant's knee-bone.[55]

Again, how pathetic. It is obvious that these people were never out there for the purposes of scientific observation. They were searching for what they <u>wanted</u> to see. And no matter what they found, they were going to force it to fit into their pre-conceived notion of evolution, even if that meant presenting the knee-bone of an elephant as *evidence* for man's descent from apes.

> "Preliminary analysis has revealed that chimp and human DNA are much more different than claimed by the evolutionary spin-doctors. Furthermore, each is rapidly accumulating mutations today. These changes in the genomes cause damage to the genome and result in birth defects, not evolution. In fact, the genomes seem to be deteriorating so rapidly that eventual extinction of both is only a matter of time."[56]

<u>- Sum Ting Wong</u> But what of Peking man or *Homo erectus*? Is this finally the missing link? (Or just a San Franciscan with good posture?)

> "Many experts consider the skulls of Peking "man" to be the remains of apes that were systematically decapitated and exploited for food by true man. Its classification, *Homo*

---

54   Brown, *In the Beginning,* 13.

55   Huse, *The Collapse of Evolution,* 100.

56   Morris, *The Young Earth,* 76.

erectus, is considered by most experts to be a category that should never have been created."

And here's another dud…

> "The first confirmed limb bones of *Homo habilis* [not to be confused with *Homo fabulous*, a close cousin of *Homo erectus*] [Oh boy, could be a lawsuit there.] were discovered in 1986. They show that this animal clearly had apelike proportions and should never have been classified as manlike (*Homo*)."[57]

"Your hands have made me and fashioned me" (Psalms 119:73)

- <u>Southwest Colorado Man</u> was another fantastic discovery that added to the *reams of evidence* (bupkis in reality) of apes morphing into men, but it was also nothing but a tooth, which turned out to be that of a horse.

How impressive and creative human evolutionists are. They can come up with a complete species of prehistoric man from a single tooth. Of a horse![58]

> "Bones of modern-looking humans have been found deep in undisturbed rocks that, according to evolution, were formed long before man began to evolve. Examples include the Calaveras skull, the Castenedolo skeletons, Reck's skeleton [not to be confused with Red Skelton, another great American patriot] and others. Remains such as the Swanscombe skull, the Steinheim fossil, and the Vertesszollos fossil presents similar problems. Evolutionists almost always ignore these remains."[59]

"See no evil, hear no evil, speak…" "Acknowledge no truth, publish no truth, teach no truth…" The mission statement of evolution instruction.

- <u>Neanderthal Man</u> was once hailed as another missing link between apes and men, but now has been re-classified as entirely human.[60]

---

57   Brown, *In the Beginning*, 13.

58   Huse, *The Collapse of Evolution*, 98.

59   Brown, *In the Beginning*, 14.

60   Huse, *The Collapse of Evolution*, 101-102.

> "For about 100 years the world was led to believe that Neanderthal man was stooped and apelike. This false idea was based upon some Neanderthals with bone disease such as arthritis and rickets. ... Neanderthal man, Heidelberg man, and Cro-Magnon man are now considered completely human. Artist's drawings of "ape-men," especially their fleshy portions, are often quite imaginative and are not supported by the evidence."[61]

Another one bites the dust. Scratch one more *transition* between apes and men. But don't give up yet! There is still hope. Anthropologists are now very excited about the possibility of studying a missing link that, incredibly, is still <u>alive</u>! ("Joe Biden, could you please report to the Smithsonian.")

"Your hands have made me... You have clothed me with skin and flesh, and have knit me together with bones and sinews." (Job 10:8, 11)

<u>- I Love Lucy</u> And finally there is *Lucy*. Not Ricky Ricardo's wife, but an *australopithecine* trumpeted by National Geographic Magazine (which has never met an evolutionary hoax that it did *not* embrace) in a December 1976 article as another missing link, and as having "walked on two legs." However, an unbiased reading of her bone fragments now suggests that Lucy was a chimpanzee. In addition, evidence has shown that prior to the time of Lucy, humans were already walking upright. (Like they ever weren't?) Apparently then we would now have to reject ol' Lucy as being an ancestor to man. Because man did not HAVE any "ancestors," and human evolution is a joke.[62]

> "The australopithecines, made famous by Louis and Mary Leakey, are quite distinct from humans. Several detailed computer studies of australopithecines have shown that their bodily proportions were not intermediate between those of man and living apes. Another study, which examined their inner ear bones, used to maintain balance, showed a striking similarity to those of chimpanzees and gorillas, but great differences from those of humans. Likewise, their pattern of dental development corresponds

---

61   Brown, *In the Beginning*, 13.

62   Huse, *The Collapse of Evolution*, 102-103.

to chimpanzees, not humans. …[Lucy] likely swung from
the trees and was similar to pygmy chimpanzees."[63]

The same is true of all the other so-called missing links between apes
and men. They are frauds. Indeed, finding a few bone fragments in the dirt
and extrapolating that as proof for the idea that apes could somehow turn
into men, is no less absurd than if you found a rusty bolt and a bent license
plate in a glacier in Alaska and hypothesized that Detroit does not exist and
that mid-size sedans somehow changed on their own, accidentally by random
chance, over many, many years, into SUV's.

"And now let us here the conclusion of the whole matter." The theory
of evolution is a joke. And a bad one at that. It is absurd. There are no
transitional forms from apes to man, because man did not come from apes.
Man proceeded forth from the command of the Almighty. As did all the
living. He spoke them all into being with the breath of his mouth. And that
is a scientific fact.

'The Spirit of God has made me, and the breath of the Almighty has
given me life." (Job 33:4)

[However, and not to contradict everything we just said, but there
have been some recent developments that have shed new light on this entire
subject. And in the interest of fairness and honesty, we would be remiss if we
did not report these facts. A new species of human-like creatures <u>has</u> been
discovered, in 2011. And they are called *Pithe-yeswecanus Leftus-Pelosius*, or
Democrat Man. The last sighting of these primitive, ape-like, pre-humans was
in Washington D.C. where they were observed destroying the U.S economy,
apologizing for the greatest country on the face of the earth, undermining our
National Defense and attempting to eviscerate the world's greatest Health
Care system. Maybe evolution isn't so crazy after all? Stay tuned for further
updates…]

So, in conclusion, there is no scientific evidence that supports the theory
of evolution, because evolution doesn't exist. It's built on lies, fabrications
and hoaxes. It truly is a joke. Now let's take a look at all of the real evidence.
(Hey! Is this fun or what?)

---

63   Brown, *In the Beginning,* 13-14.

PART II

"FOLLOW THE EVIDENCE WHEREVER IT LEADS"

"In the beginning God created the heaven and the earth… the seas, and all that in them is… And God said, Let the waters bring forth abundantly the moving creature that has life, and fowl that may fly above the earth in the open firmament of heaven. And God created great whales, and every living creature that moves, which the waters brought forth abundantly, after their kind, and every winged fowl after his kind: and God saw that it was good. …And God said, Let the earth bring forth the living creature after his kind, cattle, and creeping thing, and beast of the earth after his kind: and it was so. And God saw that it was good." (Genesis 1: 1, Exodus 20:11, Genesis 1:20-21, 24-25)

We have seen how Darwinists over the last century have taken physical evidence and, with the help of their willing accomplices in our anti-Christian, anti-Bible, anti-Creation media, educational system and entertainment industry, have distorted, lied about, fabricated and repressed it in order to further their religion-of-the-absurd. Now it's time to see what science, and not pseudo-science, has to say about all of the evidence. Let's see what happens when you look at the world without evolution-colored glasses. It all suddenly becomes very clear…

The Laws of science

"For God, who commanded the light to shine out of darkness, has shined in our hearts." (2 Corinthians 4:6)

The laws of science by themselves render the theory of evolution impossible: The Law of Biogenesis, the Second Law of Thermodynamics and the Laws of Mathematical Probability. There are others that could be included, like the First Law of Thermodynamics, but those three will suffice.

And remember that laws of science are so well-established that they never have been, and cannot be, contradicted. Otherwise they would not be laws. Theories, conversely, are literally a dime a dozen. They come and go as scientific observation either confirms or refutes them, but under any other circumstance except for when it comes to evolution, <u>no</u> theory would ever be proposed that contradicts well-established laws of science. Because the person proposing that theory would be laughed out of the scientific community. Evolution, however, gets a special exemption. You see, evolution <u>has</u> to be true. Its devotees have far too much invested in it and their professional humiliation cannot be allowed; and they are protected from widespread exposure because of their control over our media, educational system and entertainment industry. So, "Damn the truth. Full speed ahead! Evolution is a *fact*, kids, and don't ever dare to question that doctrine! Ignore that little man behind the curtain, and never forget, you're <u>stupid</u> if you think otherwise!"

<u>- The Law of Biogenesis</u>[64] is an established <u>law of science</u>. It states that life comes only from life. The theory of evolution stands in direct contradiction to this law. It says that life arose spontaneously on its own from inorganic matter by random accidental chance. Yet, in addition, according to the Laws of Mathematical Probability, that is absolutely impossible. Again, nothing with odds greater than 1 in 1066 (ten with sixty six zeroes behind it, which is one million times a trillion times a trillion times a trillion times a trillion times a trillion) can <u>ever</u> happen even if the universe were to last for an infinite number of years. And as we shall see in the pages that follow, the odds of even a tiny fractional part of a living cell making itself accidentally are far beyond 1 in 1066.

> "Spontaneous generation (the emergence of life from nonliving matter) has never been observed. [And never will be.] All observations have shown that life comes only from life. This has been observed so consistently it is called *the law of biogenesis*. The theory of evolution conflicts with this scientific law when claiming that life came from nonliving matter through natural processes."[65]

---

64   (Again, not to be confused with the so-called "Biogenic Law" which is a complete falsehood, as we saw in an earlier section, and no "Law" of science.)

65   Brown, *In the Beginning*, 5.

<u>- The Second Law</u> It is also scientifically impossible for life to have arisen on its own according to the Second Law of Thermodynamics which states that this universe and everything in it is running down, deteriorating, becoming *less* ordered. Life of course is order and complexity to the highest magnitude and for it to have arisen on its own accidentally would be in direct violation of this well-established law of science.

> "The observed laws of science provide a huge barrier to evolution. [He's being kind- they destroy it.] A basic law of science, known as the "second law of thermodynamics," has been observed and verified in every field of science. This law reveals a general trend toward deterioration in all of nature. Everything moves in a downward spiral toward a less ordered state. Stars burn out. The moon's orbit is decaying. Cars wear out. Wood rots. Living plants wither and fade. Animals die. People get sick and grow old. No exception to this rule of decline has ever been observed."[66]

*But wait! What about rock crystals? Aren't they an example of order arising spontaneously? And doesn't **that** overturn the Second Law of Thermodynamics?!* No it does not...

> "A crystal growing out of a solution... only follows well-known laws of molecular attraction and bonding and merely conforms to a repetitious pattern. There is no comparison to the much more complex patterns found in living molecules. No known law exists to order the growth of life systems."[67]

The same is true of snowflakes. They, like crystals, follow the laws of molecular attraction. But there are no laws of science or nature that could cause your camera or your computer to form itself, nor your eye or your brain. The Second Law of Thermodynamics destroys the theory of evolution all by itself.

<u>- Statistics and Probability</u> Also, both the Laws of Mathematical Probability—as we will see in the upcoming section "What are the odds?"—

---

66   Morris, *Dinosaurs, the Lost World, and You*, 7.

67   Morris, *Dinosaurs, the Lost World, and You*, 8.

and the Second Law of Thermodynamics render the idea of an organism accidentally turning into another life form by unconsciously altering its DNA as equally scientifically impossible.

> "Many molecules necessary for life, such as DNA, RNA, and proteins, are so incredibly complex that claims they evolved are questionable.

[That author is also being kind, the claims are laughable.]

> …"There is no reason to believe that mutations or any natural process could ever produce any new organs—especially those as complex as the eye, the ear, or the brain. For example, an adult human brain contains over 1014 (a hundred thousand billion) electrical connections, more than all the soldered electrical connections in the world. The human heart, a ten-ounce pump that will operate without maintenance or lubrication for about 75 years, is another engineering marvel."[68]

That's what the <u>laws</u> of science say. And next we will present a treasure trove of facts and evidence that add to their incontrovertible decree…

"Just the facts, ma'am. Just the facts."[69]

There are one hundred trillion living cells in your body. Touch your skin with the tip of your finger and you are touching millions of cells just on the surface of your skin.

> "If all the DNA in [just] one of your cells were uncoiled, connected and stretched out, it would be about seven feet long. It would be so thin its details could not be seen, even under an electron microscope. If all this very densely coded information from *one cell* of one person were written in books, it would fill a library of about 4000 books. If all the DNA *in your body* were placed end-to-end, it would stretch from here to the moon more than 500,000 times!

---

68 Brown, *In the Beginning*, 7.

69 (A catch phrase made famous by Sergeant Joe Friday in the classic TV series *Dragnet*.)

[The average distance from the earth to the moon is 240,000 miles, so all the DNA in your body would *stretch out* to 120 <u>billion</u> miles! From the DNA from JUST ONE HUMAN BODY!! And we were made by "random chance, accidentally." What an idiotic, colossal joke! Truly we are "fearfully and wonderfully made." (Psalm 139:14)]

> …"In book form, that information [from *all* of your cells] would fill the Grand Canyon almost 100 times. [Yet] If one set of DNA (one cell's worth) from every person who ever lived [approximately 50 billion people] …—enough to define the physical characteristics of all those people in microscopic detail— …were placed in a pile, [it] would weigh less than an aspirin! Understanding DNA is just one small reason for believing that you are *"fearfully and wonderfully made."* "[70]

"Understanding DNA is just one small reason for" knowing that it is a scientific fact that the theory of evolution is scientifically absurd. Don't skip over that last paragraph too quickly. Read it slowly and make sure the information there—staggering facts that are almost impossible for us to comprehend—sinks in deeply and pushes out <u>once</u> and for all any possibility of thinking that any of it could ever have evolved. The micro-miniaturization of this information, crammed into a seven-foot-strand of a molecular computer that is so microscopically thin that just the DNA in your body would stretch from here to the moon a half million times! If we humans assembled all of the engineers on the planet and took all of the cables of all of the suspension bridges in the world and tried to slice them thin enough to go from here to the moon even once we would find it nearly impossible. And all of those cables would probably fill a football stadium a number of times over. But just the DNA in just one human body, stretched out end to end, would go from here to the moon 500,000 times. Who did that? Who had the limitless power and the infinite intelligence necessary to do that? Keep in mind that we think it is incredible when a South Korean artist can carve Michelangelo's *Madonna and Child* on a rice kernel, and rightly so. But this is like taking a million IBM mainframe computers from NASA, each filled to capacity with data, and reducing them to the *tip* of a pin. And it all happened by random chance?!#@#!? Accidentally!?#@%#@?! Don't get me wrong. The obvious intention of this book is to educate people as to the truth,

---

70   Brown, *In the Beginning,* 3, 75.

and <u>not</u> to put people down, or to demean, disparage, ridicule and mock the other side as they have so delighted in doing to us over the years. But, c'mon! Let's set the record straight. Anyone at room temperature looking at this information can't help but laugh out loud at the theory of evolution. I was not exaggerating when I said that it is a joke that literally borders on the insane. And we are just getting started…

> "Cells require a remarkable array of sophisticated signal processing behaviors that rivals or surpasses that of modern computers."[71]

Question: Would any parent allow their children to be indoctrinated by some irrational human being who taught that his laptop just formed itself accidentally by random chance, and said that his students were idiots if they believed in Microsoft?

> …"the arrangement of multiple genetic elements into sophisticated logic circuits similar to those of computers is also well beyond the edge of Darwinian evolution."[72]

The DNA molecule, the information and command center of the cell, contains more coded information than a school library. And it is a scientific fact that coded information can never, ever, ever arise on its own by dumb luck, by random chance. Coded information requires a *Coder*. Now I've met Random Chance and I can tell you, he's no Coder. He cannot see, he cannot speak, he cannot think, he cannot design, he cannot create, and he definitely cannot code. In fact, he is indistinguishable from an idiot. And "those who make him into something that he is not are like unto him; and so is everyone who has faith in him." (Psalms 115:5-8) The worshipping of idols of carved stone and wood in biblical times has given way to the bowing down to the more *intellectually palatable* idols of today: Mr. Random Chance and Mr. Explosion. Otherwise known as the great gods "Chaos." They are the ones

---

71   ("Bhattacharyya, R. P. Remenyi, A., Yeh, B. J., and Lim, W. A. 2006. Domains, motifs, and scaffolds: the role of modular interactions in the evolution and wiring of cell signaling circuits. *Annu. Rev. Biochem.* 75:665-80. As quoted on pg. 270 of, *The Edge of Evolution: The Search for the Limits of Darwinism,* by Dr. Michael J. Behe)

72   Dr. Michael J. Behe, *The Edge of Evolution: The Search for the Limits of Darwinism* (New York: Free Press, 2007), 275.

who are now worshipped in place of the Creator God, who clearly warned us, "You shall have no other gods before me." Exodus 20:3)

> "A **code** *is a set of rules for converting information from one useful form to another.* Examples include Morse code and Braille. Code makers must simultaneously understand at least two ways of representing information and then establish the rules for converting from one to the other and back again. [And the idea that random chance could accomplish this is literally insane.]

> "The genetic material that controls the physical processes of life is coded information. Also coded are complex and completely different functions: the transmission, translation, correction, and duplication systems, without which the genetic material would be useless, and life would cease. It seems most reasonable that *the genetic code* and the accompanying transmission, translation, correction, and duplication systems were produced simultaneously in each living organism by an extremely high intelligence.

[Indeed, it is "most reasonable" because there is no other way in the known universe that it could be produced. Unless you know of another way that your laptop, your fridge, your toaster, your house and your car could all be produced, accidentally…]

> "Likewise, no natural process has ever been observed to produce a program. *A* **program** *is a planned sequence of steps to accomplish some goal.* Computer programs are common examples. Because programs require foresight, they are not produced by chance or natural processes. The information stored in the genetic material of all life is a complex program. Therefore, it appears that an unfathomable intelligence created these genetic programs. [Apparently.]

> "Life contains matter, energy, and **information**. All isolated systems, including living organisms, have specific, but perishable, amounts of information. No isolated system has ever been shown to increase its information content significantly. Nor do natural processes increase information; they destroy it. Only outside intelligence can significantly increase the information content of an otherwise isolated

system. All scientific observations are consistent with this generalization, which has three corollaries:

- Macroevolution cannot occur.

- Outside intelligence was involved in the creation of the universe and all forms of life.

- Life could not result from a "big bang."

…"As explained above, only intelligence creates codes, programs, and information (CP&I). Each involves senders and receivers. Senders and receivers can be people, animals, plants, organs, cells, or certain molecules. (The DNA molecule is a prolific sender.) The CP&I in a message must be understandable *and* beneficial to both sender *and* receiver; otherwise, the effort expended in transmitting and receiving messages (written, chemical, electrical, magnetic, visual, and auditory) will be wasted.

"Consider the astronomical number of links (message channels) that exist between potential senders and receivers: from the cellular level to complete organisms, from bananas to bacteria to babies, and across all of time since life began. All must have compatible understandings (CP&I) and equipment (matter and energy). Designing compatibilities of this magnitude requires one or more *super*intelligences. Furthermore, these superintelligence(s) must completely understand how matter and energy behave over time. In other words, the superintelligence(s) must have made, or at least mastered, the laws of chemistry and physics wherever senders and receivers are found. The simplest, most parsimonious way to *integrate all of life* is for there to be only one superintelligence. [Hmmm… God?]

"Also, the sending and receiving equipment, including its energy sources, must be in place and functional before communication begins. But the preexisting equipment provides no benefit until useful messages begin arriving. Therefore, intelligent foresight (planning) is mandatory— something nature cannot do."[73]

Isn't knowledge a wonderful thing? And with that in mind we'd like to offer a group discount to Richard Dawkins, Christopher Hitchens, Bill

---

73   Brown, *In the Beginning,* 9.

Maher and Sam Harris if the four of them sign up together for the next class on *Remedial Re-education 101: Basics in Biology*. Shucks, we'll even throw in a second one for free. It's called, *If you feel compelled to be a completely ignorant, arrogant, abusive, intolerant, militant and thoroughly misinformed atheist it might help to actually know what the hell you are talking about first, 201*. David Berlinski had this observation…

> "When asked what he was in awe of, Christopher Hitchens responded that his definition of an educated person is that you have some idea of how ignorant you are. This seems very much as if Hitchens were in awe of his own ignorance, in which case he surely has found an object worthy of his veneration."[74]

But let us get back to the reasons for being in awe of the Creator of this universe, rather than in our own ability to deny Him and to hide that denial behind an idiotic and unscientific theory…

> "The notion that not only the biopolymers but the operating programme of a living cell could be arrived at by chance in a primordial organic soup here on Earth is evidently nonsense of a high order."[75]

But "nonsense of a high order" that is nevertheless taught, as fact, with a straight face, every day to the public school children all across America. And we wonder why our country is going to hell in a handbasket, and why our voting-age youth have no more sense than to rush out and vote for the left-wing Democrat jackasses that they've been told to vote for. Truly sad. Public indoctrination, and brainwashing.

> "…consider the information in the DNA molecule. No undirected process produces intelligent information that can be read and understood. Its coded information surpasses our own ability to understand it. Today's molecular biologists cannot write such a code or even devise a way to do the things a cell can do. Certainly a level

---

74   David Berlinski, *The Devil's Delusion: Atheism and its Scientific Pretensions,* (New York: Crown Publishing Group, 2008), 208.

75   (British astronomer, Sir Fred Hoyle)

of intelligence beyond our own was behind the writing of the DNA [and] the functions of a cell."[76]

Mutations never result in an increase of information in the DNA code; they always result in a decrease. Mostly they result in horrible defects and distortions and cause death. Therefore, from this <u>observation</u> alone (and remember that science is based on systematic observation) the evolutionists belief that one organism (i.e. an ape) can change into another (i.e. a man) by random accidental genetic mutation over whatever period of time they would care to fabricate, is completely false. *Everyone* knows that it is scientifically impossible for your computer, your camera, your tape recorder, your automobile, the wiring in your house, your smoke detector, our Star Wars Defense System, your refrigerator or your cars fuel system to have arisen accidentally over time. If anyone said otherwise they would be called a moron. Well it is even more scientifically impossible (according to mathematical probabilities) for your brain, your eyes, your ears, your skeletal system and muscles, your nerves, your nose, your immune system, your sweat glands or your digestive system to have arisen accidentally over time. And if anyone says otherwise then by extension they would have to be a moron as well, wouldn't they? (Or at least someone who was duped into believing the nonsense/non-science they were taught.)

Our DNA computer in each of our cells tells us what kind of eyes, skin, hair and height we will have. It also programs the organization of all 600 of our muscles, 10,000 nerves for hearing, a couple million optical nerves, 60 thousand miles of capillaries and blood vessels (see below), and over 200 bones. The ability of the tiny DNA strand (there are 100 trillion of them in your body, one inside of each of your cells) for information storage totally dwarfs that of our modern computers. All of the DNA of every species that is alive now or has ever lived, needed to determine all of their characteristics and design could fit in the palm of a baby's hand, and you would still have space for all of the words in every book ever written. And yet our evolutionary idiots in our colleges, universities, elementary and high schools laugh out loud at the very mention of the idea of an intelligent designer. They are of course laughing at themselves, as they brainwash our unsuspecting, naive and gullible children. They are, of course, literally insane.[77]

---

76   Morris, *The Young Earth,* 123.

77   Huse, *The Collapse of Evolution,* 94.

[The theory of evolution can be totally destroyed in just three words. And what are those three words? …Sixty thousand miles. There are approximately 60,000 miles of blood vessels (veins, arteries and microscopic capillaries) in your body. The idea that they could have all put themselves there with far, far beyond micrometer accuracy so that they touch all of the 100 trillion cells in your body, accidentally, by random chance, is literally insane. After knowing this fact, you don't have to be a moron, or Ocrazyo Cortez, or a lobotomy victim to still believe in evolution, but it helps. And that is just one of thousands of such biological facts that each alone destroys evolution…]

According to computer scientists there is no way that information can happen spontaneously. It has to be directed by some intelligence to put together the letters and the words. One microscopic DNA strand contains an astronomical amount of information. Therefore random, accidental evolution doesn't exist. (Except for in the minds of evolutionists, and their poor, unfortunate, duped students.) Intelligent design is a scientific fact.

The DNA molecule is so intricately designed and engineered that it truly boggles the mind. It could never ever have just "made itself." It had to be made.

The species across this planet display an astronomically complex degree of order, and beauty, and diversity. The ability of migration, of parasites, symbiosis and mimicry that require an interrelationship between species also laugh at the idea of blind, accidental evolution, and cries out for a super intelligent and all-powerful designer.[78]

That is, of course, if one has the unfettered and un-brainwashed use of his or her mind. Remember the lies of Satan cover this earth like a blanket, and people believe whatever they are taught. (And did I mention that we live in an insane asylum, and the inmates have taken over?)

DNA obviously is necessary for life. Without it a cell cannot exist. Just as a human being cannot exist without a brain. (Well, that's debatable. Where's my ear scope? Mr. Dawkins, could you come over here and tilt your head for a second?) **Yet DNA cannot function without the presence within the cell of at least 75 pre-existing proteins**. *Yet proteins are ONLY produced by DNA!!* And neither can exist without the cell wall encasing them and surrounding them with all of the cellular chemicals and enzymes and buffers and tens of thousands of nanobot transportation robots, etc. etc. So even if a DNA molecule formed itself somehow by random chance, which

---

78   Huse, *The Collapse of Evolution*, 95.

by the way is impossible—about the same chances as the contents of The Library of Congress forming themselves by random chance—but even if it did it wouldn't do any good because it would immediately fall apart into slop. The cell wall is a miracle of engineering. It is far more technologically advanced than even our most secret and highly protected military installations or sophisticated anti-missile defense systems (and unlike those comparatively crude mechanisms, the cell wall operates completely UN-consciously). But in order for evolution to be viable not only would the DNA molecule have to, *POOF*, form itself miraculously by accident, but 75 separate proteins, and the cell wall, and all the other multitudinous contents of the cell would have to also form themselves miraculously by accident, at <u>exactly</u> the same time and at <u>exactly</u> the same place! This, of course, "would be indistinguishable from a <u>miracle</u>!" But evolution is not in the business of miracles. Only God, the Creator, is. Evolution is in the non-existent, imaginary business of small, slow, incremental, mythological changes over vast spans of time, which we will see shortly, also don't exist.

"It's evolution that doesn't exist, stupid, not God."

> "The complexity of the simplest known type of cell is so great that it is impossible to accept that such an object could have been thrown together suddenly by some freakish, vastly improbable, event. **Such an occurrence would be indistinguishable from a miracle.**"
>
> "<u>Alongside the level of ingenuity and complexity exhibited by the molecular machinery of life, even our most advanced artifacts appear clumsy.</u>"[79]

"Alongside the level of ingenuity and complexity exhibited by the molecular machinery of life, even our most advanced artifacts appear clumsy." Yet no one would ever dare be so stupid as to suggest that our clumsy artifacts could just make themselves, if we waited around long enough.

Let's look even further inside the engineering miracle of a living cell, the building block of all of life…

Again, there are a hundred trillion cells in the human body. Every second there are a million chemical reactions within just one of these many cells. That's a <u>hundred million **trillion**</u> chemical reactions <u>every second</u> just

---

79   Michael Denton, *Evolution: A Theory in Crisis,* (Chevy Chase, MD: Adler and Adler, 1986), 264.

to keep your body alive and functioning. If you started right now to write down each of those chemical reactions on paper, and you did not stop day or night until you were done, it would take you well over a trillion years to do so. And that's just the chemical reactions that occur in one second within one human body.

> "Although serendipity certainly plays its part in nature, advancing sheer chance as an explanation for profoundly functional features of life strikes me as akin to abandoning reason altogether."[80]

Again, not only according to common sense but according to the *Second Law of Thermodynamics* and the *Laws of Mathematical Probability* sheer chance can <u>never</u> create anything of order. All random chance can do is to tear down, to break apart and to destroy, and you are left with a colossal mess and not anything that resembles even the simplest living thing.

Also, the actual engineering complexity of what was once called the "simple single cell" out of primitive, scientific ignorance, is greater than that of the entire city of New York. And unlike the joke that is evolution, that actually <u>is</u> a scientific fact. In each cell there are thousands upon thousands of nanobots, micro machines, tiny robots that scurry about all over the cell transporting materials like oxygen and carbon and water from the cell wall to countless places where they are needed, and taking waste products of metabolism like carbon dioxide back to the cell wall for expulsion. They do all of this over and over again millions upon billions upon trillions of times, day in and day out for your entire lifetime. And the miraculous thing is that they do it <u>blindly</u> and <u>unconsciously</u>. Everything in the cell is completely automated.

> "Cells are robots. Or rather, because they are so small, "nanobots." They work by unconscious, automatic mechanisms. To perform the routine tasks of their microscopic lives, cellular nanobots need sophisticated molecular machinery that works without conscious guidance."[81]

Can you imagine constructing an entire city of New York that could run itself for decades blindly and unconsciously, completely automated, with

---

80   Behe, *The Edge of Evolution*, 165.

81   Behe, *The Edge of Evolution*, 19.

no human intervention? A car gets in an accident, the tow trucks would have to be able to sense it automatically and drive to the right place and pick it up and then bring it to the garage where it would be repaired by robots working blindly and unconsciously without any human intervention. The same with a water main that breaks; a fire that starts; a hurricane that blows the roofs off and the windows out of countless buildings; an elevator or air conditioning or heating system that breaks down. The repair robots would have to know exactly where the water leak is, the fire is, the windows are blown out, blindly and unconsciously! Think about it! If we gathered all of the world's smartest engineers and scientists and allowed them to work day and night for the next ten million years, they could never do it. After a few million years they might get a small scale primitive prototype, consisting of maybe ten square blocks, up and running and it might last for maybe half a day until something blew up that the automated city could not handle on its own and the humans would have to rush in and take over. Back to the drawing board. But the Intelligent Creator of this universe accomplished the equivalent of designing and manufacturing an automated city of New York within every cell of your body, and every cell of every living thing on this planet, and it took him just a few days. He spoke it, and all matter instantly obeyed his voice. He breathed it, and it all sprang into being. He wound it all up like a clock, and it's been running itself, unconsciously, ever since.

> "Most people don't think of cells as robots, probably because cells are made of organic materials rather than metal. But cells truly are self-replicating nanoscale robots. Self-replicating, because of course they reproduce themselves. Nanoscale, because most cells are quite tiny and all can manipulate single molecules. Robots, because their activities are carried out unconsciously and automatically by precision machinery that follows ordinary physical laws."[82]

> "One crucial way in which machinery in the nanobot differs from machinery in our everyday experience is that cellular machines have to *assemble themselves*. There is no conscious agent walking around in the cell putting pieces of machinery together, as there might be in a factory making, say, flashlights. Needless to say, the requirement for self-

---

82   Behe, *The Edge of Evolution*, 242.

assembly enormously complicates the task of building a functional nanomachine."[83]

"Needless to say" we are talking about a level of astronomical complexity and supreme design ingenuity that absolutely dwarfs anything man has ever even conceived, much less ever manufactured or created. The "level of knuckle dragging ignorance" (throwing back in his face the famous atheist and evolutionist Mr. Richard Dawkin's famous false accusation against those who fail to agree with his blind faith in the complete idiocy that is evolution, and atheism)… "the level of knuckle dragging ignorance" it takes to suppose that the astronomically complex, engineering miracle of a living cell (fraught with highly advanced, completely unconscious and automated nanobot technology that we are literally thousands of years away from even fully comprehending, much less ever duplicating) could form itself accidentally by random chance is literally <u>insane</u>. And so is the theory of evolution.

…An aside: [It is spring, and I'm sitting outside looking at all of the life splashed across my backyard. And I'm thinking about the near-infinite number of chemical reactions that are occurring constantly right before my eyes in every cell of every blade of grass and weed and flower and shrub and tree, and all of the nearly as infinite number of microscopic little robots that are unconsciously scurrying around each cell, and all of this is happening just so we can experience life. Life is happening microscopically and miraculously in its infinite and rigidly ordered astronomical complexity in order to bring us: life. And yet it all happened by random, accidental chance?!#%&*! What an ignorant slap in the face to an incredible and unfathomably intelligent and infinitely powerful and loving and marvelous Creator who has blessed us with this perpetual miracle that he has showered across the face of this earth. It's a wonder he doesn't just stomp us out.

The wonder of life is far beyond our ability to even comprehend it, much less fully appreciate it. It is so true that "you don't know what you got till it's gone." Pray for all of those who are mis-educated into the false-religious lies of the great deceiver, who stand to lose this miracle of life for all eternity, never again to have the joy of experiencing it.] [See chapter 7 of All the Secrets.]

---

83   Behe, *The Edge of Evolution*, 255.

We will leave the incredibly-complex, molecular-sized world of the living cell now and look at the evidence from other perspectives. Again, "just the facts, ma'am, just the facts"…

I bet you never thought that even altruism could refute evolution. Neither did I…

> "Humans and many animals will endanger or even sacrifice their lives to save another—sometimes the life of another species. Natural selection, which evolutionists say select individual characteristics, should rapidly eliminate altruistic (self-sacrificing) "individuals." How could such risky, costly behavior ever be inherited? Its possession tends to prevent the altruistic "individual" from passing on his genes for altruism? If evolution were correct, selfish behavior should have completely eliminated unselfish behavior. Furthermore, cheating and aggression should have "weeded out" cooperation. Altruism contradicts evolution."[84] [Life contradicts evolution, but let's not pile on.] [And see page 8 of *In The Beginning* for further discussion and examples.]

And then there is language…

> "Children as young as seven months can understand and learn grammatical rules. Furthermore, studies of 36 documented cases of children raised without human contact (feral children) show that language is learned only from other humans; humans do not automatically speak. So, the first humans must have been endowed with a language ability.
>
> …"If language evolved, the earliest languages should be the simplest. But language studies show that the more ancient the language (for example: Latin, 200 B.C.; Greek, 800 B.C.; Linear B., 1200 B.C.; and Vedic Sanskrit, 1500 B.C.), the more complex it is with respect to syntax, case, gender, mood, voice, tense, verb form, and inflection. The best evidence shows that languages devolve; that is, they become simpler instead of more complex. Most linguists

---

84   Brown, *In the Beginning,* 8.

reject the idea that simple languages evolve into complex languages.

...."If humans evolved, then so did language. All available evidence indicates that language did not evolve, so humans probably did not evolve either."[85]

Probably.

> "Speech is uniquely human. Humans have both a "prewired" brain capable of learning and conveying abstract ideas, and the physical anatomy (mouth, throat, tongue, larynx, etc.) to produce a wide range of sounds. Only a few animals can approximate some human sounds. Because the human larynx is low in the neck, a long air column lies above the vocal cords. This helps make vowel sounds. Apes cannot make clear vowel sounds, because they lack this long air column. The back of the human tongue, extending deep into the neck, modulates the air flow to produce consonant sounds. Apes have flat, horizontal tongues, incapable of making consonant sounds.

> "Even if an Ape could evolve all the physical equipment for speech [which he of course could not], that equipment would be useless without a "prewired" brain for learning language skills, especially grammar and vocabulary."[86]

But don't worry, this little annoying fact hasn't deterred evolutionists in the least. In fact, as we speak, they are working feverishly on a new theory of how human speech somehow *evolved* from the grunts of apes, and they're calling it *kreehayshun evolution*. It's where evolution skips all of those little annoying, incremental steps that take so long and are so scientifically-impossible, and just *evolves* any organ, any organism and any ability it wants, instantly! Shazaam! But of course no higher power or outside intelligence is involved in any way. Don't be silly.

And then there is this vaudevillian sleight-of-hand disguised as another brilliant Darwinian *explanation*...

---

85   Brown, *In the Beginning*, 8.

86   Brown, *In the Beginning*, 8-9.

"When the same complex capability is found in related organisms, but not in their alleged evolutionary ancestors, evolutionists say that a common need caused identical complexities to evolve. They call this *convergent evolution*. [It's also called *horseshit-on-a-stick*, but hey! let's try to keep the vulgarity to a minimum?]

"For example, wings and flight occur in some birds, insects, and mammals (bats). Pterosaurs, an extinct reptile, also had wings and could fly. These capabilities have not been found in any of their alleged common ancestors. Other examples of convergent evolution are the three tiny bones in the ears of mammals: the stapes, incus, and malleus. Their complex arrangement and precise fit give mammals the unique ability to hear a wide range of sounds. Evolutionists say that those bones evolved from bones in a reptiles jaw. If so, the process must have occurred at least twice—but left no known transitional fossils. How did the transitional organisms between reptiles and mammals hear during those millions of years? [Uh, hearing implants?] Without the ability to hear, survival—and reptile-to-mammal evolution—would cease.

"Concluding that a miracle—or any extremely unlikely event—happened once requires strong evidence or faith; claiming that a similar "miracle" happened repeatedly requires either incredible blind faith [called *willful ignorance*] or a cause common to each event, such as a common designer."[87] HELL-O!

Common ancestry is an evolutionary lie. Common design is a scientific truth. Of course, convergent evolution is downright hilarious. How they can even keep a straight face at the lectern while speaking it into their microphones is beyond me. By the way, in case you're not sure, it's OK to laugh out loud. Not at my lame attempts at humor, but at evolution. It really *is* a joke.

But here are even more problems for Darwinists:

"Many single-celled forms of life exist, but no known forms of animal life have 2, 3, 4, or 5 cells. Known forms of life with 6-20 cells are parasites, so they must have a

---

87   Brown, *In the Beginning,* 9-10.

complex animal as a host to provide such functions as respiration and digestion. If macroevolution happened, one should find many transitional forms of life with 2-20 cells--filling the gap between one-celled and many-celled organisms."[88]

No. Not necessarily. Not if one understands *hippety-hop evolution.* Closely related to *kreehayshun* evolution and similar to *horsh…*er…*convergent* evolution, this variation to standard Darwinism calls for the convenient leaping over of those potential 2- to 20-celled organisms not found either alive or in the fossil record. You see, when life first evolved it didn't have a calculator (*that* was the problem), so it is possible that it just missed a couple of simple numbers, like 2, 3 and 4 for example; numbers that *we* all take for granted.

Oh, but it gets even better! Much, much better…

Feeding mostly on sea anemones, the sea slug dwells in tidal areas along the ocean coast. But these anemones have tentacles that are covered with many tiny stinging cells and if they are touched they explode and paralyze the trespasser by spearing him with poisoned darts. The poor victim is then pulled inside the anemone and dissolved.

However, the incredible sea slug is not deterred by these poisonous stinging harpoons. In fact, the sea slug is able to neutralize the poison while at the same time not exploding the stinging cells while eating and digesting the sea anemone. But even more fascinating is that the stinging cells are not digested but swept into sacks where they are kept until the sea slug itself is attacked and then uses them for its own defense!!

Now, and with all due respect, the level of utter ignorance and stupidity it takes to believe and to teach that that level of astronomical complexity and sheer creativity could ever have just made itself accidentally over hundreds of millions of years is quite stunning. The sea slug would have to evolve some sort of means chemically to stop the stinging cells from exploding. But what good is that if it doesn't at the exact same time instantaneously "evolve" the digestive ability to dissolve the anemone while at the same time separating and then storing the stinging cells into pouches that it would also have to "evolve." And all at the same time.

Clearly, and from this one remarkably adapted organism alone, forgetting all of the thousands of other examples of adaptation that totally

---

88   Brown, *In the Beginning,* 10.

defy any evolutionary explanation, evolution is a joke. It doesn't exist in reality. Science itself has made a laughing stock of it. End of story.[89]

"O the depth of the riches both of the wisdom and knowledge of God! How unsearchable are his judgments and his ways past finding out!" (Romans 11:33)

> "The Arctic Tern, a bird of average size, navigates across oceans... with the skill normally associated with navigational equipment in modern intercontinental aircraft. A round-trip for the tern might be 22,000 miles. The tern's "electronics" are highly miniaturized, extremely reliable, maintenance free, and easily reproduced. Furthermore, this remarkable bird needs no training. If the equipment in [an intercontinental aircraft] could not have evolved, how could the terns more amazing equipment have evolved?"[90]

Really. How, Mr. Maher? You arrogant, uneducated ass. (With all due respect of course.) Please tell us how. As you and your imbecilic friends laugh at all of us who are just not *sharp* enough, like you, to believe in the scientifically-impossible and clearly idiotic fairy tale called evolution; and as you deride us for being "stupid, ignorant, religious and *unscientific*," please explain to us how. (Here's a crowbar in case you'd like to pry your head out of your ass first.)

And the beat goes on. (We're not even warmed up yet.) The avalanche of scientific facts that bury Darwinism for good, continues to thunder down the mountain of life...

As we quoted earlier "an adult human brain contains over 1014 (a hundred thousand billion) electrical connections, more than all the soldered electrical connections in the world."[91] Yet the human brain just formed itself by random accidental chance mutations over many years? Could all of the electrical appliances on this planet have made themselves if we just waited around long enough? Could all of the electrical wiring in the city of New York just happen by random chance if we waited long enough? Even trillions of years? Of course not. You would have to literally be insane to actually think

---

89   Huse, *The Collapse of Evolution*, 74-75.

90   Brown, *In the Beginning*, 21.

91   Brown, *In the Beginning*, 7.

that, wouldn't you? (I'm sorry, you'll have to excuse me for a minute. I can see there are some people in white coats over at Sam Harris' front door. I'd better go over and see what all the fuss is about…)

OK, I'm back…

> "The electric eel can pack an electrical punch greater than the shock that you would get from sticking your finger into an electrical socket. The electricity from this amazing creature arises from a group of highly compacted nerve endings found all along its body. Each one of these nerve endings has a small electric voltage that, when added together, can be a shocking experience! A full-grown eel can produce enough electricity—600 volts—to stun a horse. Eels use this electrical current as self-defense, and a way to stun their prey. Yet, they live in water, don't shock themselves, and can recharge without an extension cord.

[And they were created by Mr. Accidental Mutation and Mrs. Random Chance! (They got together in the back of his van after downing one too many drinks at the local pub.)]

> "Lightning bugs (also called fireflies) are little beetles that carry their own "flashlights." These insects have a special chemical called luciferin, which they store in their abdomens. When this chemical comes into contact with oxygen from the air they breathe, fireflies give off a bright flash of light. The whole process is called bioluminescence. According to scientists, this light helps fireflies find their mates. The male flashes his light first, and when the female sees it, she flashes her light. This tells the male where she is. What an incredible way to find a mate!

> "Perhaps one of the most amazing animals on Earth today is a little insect known as the bombardier beetle. This bug packs a powerful, explosive defense mechanism. Inside its body, it has tiny glands that secrete two chemicals into a "storage tank" (known as a "collecting vesicle"). Those chemicals are hydrogen peroxide and hydroquinones. If an enemy attacks the beetle, it empties these two chemicals into a special "explosion chamber." Then it adds special enzymes to the mixture. As a result, the chemicals form a solution that reaches the boiling point of water (100°Celsius

or 212°Fahrenheit) in just a couple of seconds. The beetle can then take aim with two small gun-like projections on the rear of its body, and fire the boiling mixture into the face of its attacker. What a remarkable creature!"[92]

People believe what they are taught. No matter how idiotic, asinine, absurd or ridiculous it may be (don't forget the "72 virgins" thing). And it should be becoming painfully obvious, just from the last three examples alone, that the theory of evolution falls into each of those four categories.

And then there is the woodpecker:

"A woodpecker's chisel-tipped bill hammers wood at the rate of 16 times a second, or nearly 1,000 pecking blows per minute—a "rate of fire" doubly fast as a submachine gun—with an impact velocity of 1,300 mph.

"While drilling, the woodpecker's head travels at more than twice the speed of a discharged bullet. At this speed (over 1900 feet/second), any slight whiplash rotation of the head during drilling would tear away the bird's brain upon impact. To prevent this, superbly coordinated neck muscles keep the head in perfect alignment with the beak during impact. The head and beak thus drive straight back and forth with no side movement at all. Shock-impact is further minimized by special muscles in the head which pull the woodpecker's braincase away from its beak every time it strikes a blow.

"Unlike most birds, which have their bill fused directly to the bones of the cranium, the woodpecker's bone-reinforced skull is physically separated from its beak by a remarkable sponge-like cartilage (recognized by scientists as being better than any shock absorber manufactured by man). This padding is essential to the bird's survival when one considers that the suddenness with which the woodpecker's head is brought to a halt during each peck results in a stress equivalent of 1,000 times the force of gravity. The head thus snaps back with an impact of deceleration more than 250 times that of the G-force

---

92   Butt and Lyons, *Dinosaurs Unleashed,* 62-63.

gyrations experienced by the astronauts during a launch-pad liftoff!"[93]

How did all of that slowly evolve by small incremental steps? What, as the beak of the bird slowly got stronger, the brain slowly detached from it and slowly grew that shock-absorber cartilage that we cannot duplicate, and slowly grew those "special muscles in the head" and those "superbly coordinated neck muscles," and this process slowly repeated itself thousands of times over and over, in millions and millions of years' worth of birdie generations until we finally accidentally ended up with this absolute miracle of engineering and design that our greatest scientists and most ingenious minds could not duplicate in a thousand years? Incrementally. Slowly. Accidentally. Hmm.

There are so many more examples of biological design that defy any evolutionary explanation; indeed they are too numerous to list here. How does evolution explain those creatures? How did they *evolve* those incredibly complex and mind-boggling features that we humans can't even fully comprehend much less duplicate? How could accidental mutation and random chance design and create those astonishing wonders of nature? Answers: 1) They can't explain them. All they can do is lie and distort and hide and deceive and fabricate and then refuse to ever talk about it or debate it. 2) They didn't *evolve* them. God designed and created them. 3) They couldn't. Accidental mutation and random chance can't design or create a damn thing. Like the idols of wood, stone and statues of gold worshipped by ignorant people of old; *gods* that couldn't talk or think or hear or act, much less answer a single prayer or request; so too the *gods* of accidental mutation and random chance worshipped by faux scientists, ignorant professors and their unfortunately duped students today, couldn't design and create a single, stupid lead pencil much less the wonders of God's nature.

And then there is metamorphosis. To attempt to explain that with Darwinism is like trying to explain advanced calculus with the farts of buffalos:

> "Most insects (87%) undergo complete metamor-phosis. It begins when a larva (such as a caterpillar) builds a cocoon around itself. Then its body inside disintegrates

---

93   (From *The Wonderful Woodpecker: Jehovah's Jaw-Jarring Jackhammer*, by David V. Bassett, M.S. posted on astronomy.net)

into a thick, pulplike liquid. Days, weeks, or months later, the adult insect emerges—one that is dramatically different ...amazingly capable, and often beautiful, such as a butterfly. Food, habitat, and behavior of the larva also differ drastically from those of the adult.

...."The millions of changes inside the thick liquid never produce something survivable or advantageous in the outside world until the adult completely forms. How did the genetic material for both larva and adult develop? Which came first, larva or adult? What mutations could transform a crawling larva into a flying monarch butterfly that can accurately navigate 3,000 miles using a brain the size of a pin head? Indeed, why should a larva evolve in the first place, because it cannot reproduce?

"Charles Darwin wrote, *If it can be demonstrated that any complex organ existed which could not possibly have been formed by numerous successive, slight modifications, my theory would absolutely break down.* Based on metamorphosis alone, evolution "breaks down."

"Obviously, the vast amount of information that directs every stage of a larva's and adult's development, including metamorphosis, must reside in its genetic material at the beginning. This fits only creation."[94]

And then there are numerous examples of symbiotic relationships between living things, each dependent upon the other for their survival, that are still waiting for an intelligent and scientifically viable evolutionary explanation. The silence has been deafening for decades now. (Mr. Gutfeld, we're still awaiting with bated breath. Greg, are you there? Tyrus? Come out, come out, wherever you are.)

"Different forms of life are completely dependent upon each other. At the broadest level, the animal kingdom depends on the oxygen produced by the plant kingdom. Plants, in turn, depend on the carbon dioxide produced by the animal kingdom.

"More local and specific examples include fig trees and the fig gall wasp, the yucca plant and the yucca

---

94   Brown, *In the Beginning,* 17-18.

moth, many parasites and their hosts, and pollen-bearing plants and the honeybee. Even members of the honeybee family, consisting of the queen, workers, and drones, are interdependent. If one member of each interdependent group evolved first (such as the plant before the animal, or one member of the honeybee family before the others), it could not have survived. Because all members of the group obviously have survived, they must have come into existence at essentially the same time. In other words, creation."[95]

And then there is the whole idea of sex. Evolutionary explanations are incomprehensible.

"If sexual reproduction in plants, animals, and humans is a result of evolutionary sequences, an unbelievable series of chance events must have occurred at each stage.

a. The amazingly complex, radically different, yet complementary reproductive systems of the male and female must have **completely** and **independently** evolved at each stage at about the **same time** and **place**. [Of course, that is a scientific impossibility.] Just a slight incompleteness in only one of the two would make both reproductive systems useless, and the organism will become extinct.

b. The physical, chemical, and emotional systems of the male and female would also need to be compatible.

c. The millions of complex products of a male reproductive system (pollen or sperm) must have an affinity for and a mechanical, chemical, and electrical compatibility with the eggs of the female reproductive system.

d. The many intricate processes occurring at the molecular level inside the fertilized egg would have to work with fantastic precision—processes that scientists can describe only in a general sense.

e. The environment of this fertilized egg, from conception through adulthood and until it also reproduced with another sexually capable adult (who also "accidentally" evolved), would have to be tightly controlled.

---

95   Brown, *In the Beginning,* 18.

f. This remarkable string of "accidents" must have been repeated for millions of species.

Finally, to produce the first life form would be one miracle. But for natural processes to produce life that could reproduce itself would be a miracle on top of a miracle."[96]

And it is! It's called Creation…

"For in six days the Lord made heaven and earth, the sea, and all that in them is, and rested the seventh day: wherefore the Lord blessed the Sabbath day, and hallowed it." (Exodus 20:11)

Next we have the immune system:

"How could immune systems of animals and plants have evolved? Each immune system can recognize invading bacteria, viruses, and toxins. Each system can quickly mobilize the best defenders to search out and destroy these invaders. [And they do that automatically and without conscious intervention, thousands of trillions of times, over and over again, every minute of every day.] Each system has a memory and learns from every attack.

"If the many instructions that direct an animal's or plant's immune system had not been pre-programmed in the organism's genetic system when it first appeared on earth, the first of thousands of potential infections would have killed the organism. This would have nullified any rare genetic improvements that might have accumulated. In other words, the large amount of genetic information governing the immune system could not have accumulated in a slow, evolutionary sense. Obviously, for each organism to have survived, all this information must have been there from the beginning. Again, creation."[97]

"It's impossible. Ask a baby not to cry. It's just impossible. Evolution is a lie. It's so impossible."[98]

---

96   Brown, *In the Beginning,* 19.

97   Brown, *In the Beginning,* 19.

98   (Sung to the tune of *It's Impossible,* song and lyrics by Perry Como)

How many times have you found yourself watching a science channel on your TV and you hear something like this example from the National Geographic Channel, "The Great White Shark glides through the sea with minimum effort and perfect grace. A shape of ideal hydrodynamic **design**." [emphasis mine] And it is such an obvious oxymoron for those evolutionists to <u>ever</u> be using that word. I've heard that word "design" used hundreds of times for the snakes in the deserts and the sheep on the mountains, for the gophers in the prairie and the foxes and polar bears in the snow-covered tundra of Alaska, and I find myself shouting at the TV, "So, who #*&#*@% *designed* them then, you knuckleheads?!" Anyone with a brain and some common sense knows that design requires a **designer**! And that Great White Shark is more beautifully and precisely designed for its purpose than a flipping Lamborghini. But they just don't get it, because they love the lie more than they love their own Creator (and Designer). So He gives them over to a reprobate and illogical mind. One that is "ever learning, yet never able to come to the knowledge of the truth." And one that has been imprisoned by the asinine, idiotic, scientifically-impossible, bald-faced lie of evolution, Satan's genesis.

> "Most complex phenomena known to science are found in living systems—including those involving electrical, acoustical, mechanical, chemical, and optical phenomena. Detailed studies of various animals also have revealed certain physical equipment and capabilities that the world's best designers, using the most sophisticated technologies, cannot duplicate. [And that, of course, is because the One who designed these living systems is infinitely more intelligent than any human designer.] Examples of these designs include molecular-size motors in most living organisms; advanced technologies in cells; miniature and reliable sonar systems of dolphins, porpoises, and whales; frequency-modulated "radar" and discrimination systems of bats; efficient aerodynamic capabilities of hummingbirds; control systems, internal ballistics, and the combustion chambers of bombardier beetles; precise and redundant navigational systems of many birds, fish, and insects; and especially the self-repair capabilities of almost all forms of life. No component of these complex systems could have evolved without placing the organism at a selected disadvantage until the

component's evolution was complete. All evidence points
to intelligent design.

[But our left-wing, lying media, entertainment industry and educational
system will continue to shout the fiction that those who believe in intelligent
design are *ignorant, unscientific fundamentalists*, and those who believe in
evolution are the "rational" "scientific" ones. What a hoot! What a lie!] [Kind
of like their "Democrats smart! Conservatives stupid! See Spot run" myth.]
[But for an in depth expose of that subject, see chapter 13 "Satan, The Great
Deceiver" in the larger work, "All the Secrets..."]

> "Many bacteria, such as *Salmonella, Escherichia coli,*
> and some *Streptococci,* propel themselves with miniature
> motors at up to 15 body-lengths per second, equivalent
> to a car traveling 150 miles per hour—in a liquid. These
> extremely efficient, reversible motors rotate at up to
> 100,000 revolutions per minute. Each shaft rotates a bundle
> of whiplike flagella that acts as a propeller. The motors,
> having rotors and stators, are similar in many respects to
> electrical motors.

[If someone actually believed that your refrigerator motor just evolved
accidentally by random chance, you would call them a moron. Yet our
naïve, gullible and unsuspecting school children in our colleges, universities,
elementary and high schools are taught by complete morons every day. And
we all sit back like equal morons and let them continue unexposed....]

> However, their electrical charges come from a flow
> of protons, not electrons. The bacteria can stop, start,
> and change speed, direction, and even the "propeller's"
> shape. They also have intricate sensors, switches, control
> mechanisms, and a short-term memory. All this is highly
> miniaturized. **Eight million** of these bacterial motors
> would fit inside the circular cross section of a human hair.

[And yet rather than allowing our students to revel in the awesome,
mind-boggling imagination and creative power of their Maker, they shovel
a bald-faced, scientifically-absurd pile of horseshit down their throats: the

great deceiver's "origins" account. And their uninformed parents just sit back and let it happen.[99] How sad. How very sad.]

> "Evolutionary theory teaches that bacteria were one of the first forms of life to evolve, and, therefore, they are simple. While bacteria are small, they are not simple. They can even communicate among themselves using chemicals.
>
> "Some plants have motors that are one-fifth the size of bacterial motors. Increasing worldwide interest in nanotechnology is showing that living things are remarkably designed—beyond anything Darwin could have imagined."[100]

Darwin had an excuse. He could not possibly have imagined it. Today we have no excuse. Only willful ignorance.

"The hearing ear and the seeing eye, The Lord has made them both." (Proverbs 20:12)

How about a movie camera? Could the universe make a movie camera, by random chance, starting from rocks and dirt, using small incremental accidental steps? Of course not. You'd have to be a moron to think so. Well how about the human eye? Could the universe make a human eye, by random chance, starting from a mixture of chemicals and water, using small incremental accidental steps? Of course not. You'd have to believe the ridiculous lies that you were taught in order to think so. And keep in mind that the human eye is literally hundreds of thousands of times more complicated than a movie camera. First of all, the human eye forms itself in the womb blindly and unconsciously from just the DNA blueprints contained inside of each cell. It performs this incredible task with no conscious intervention, all completely automated, millions of tiny little molecular machines running around forming each specific type of cell and groups of cells and putting countless billions of them in precisely the right spot, building your eye in the womb, using the coded information from a tiny strand of DNA. And then your eye will go on to operate itself 24/7 without any conscious intervention for the rest of your life. It is a completely automated system that operates constantly, repairs itself, and captures and interprets trillions of pixels of visual

---

99   (And then rush out and vote for the baby-killing Democrats. Yea, I just had to throw that in)

100   Brown, *In the Beginning*, 19.

information every picosecond, transmitting it all reliably and accurately to your brain; all accomplished since the beginning of life on earth over and over again, trillions upon trillions of times, in every organism with sight capabilities; again, completely automated with no conscious intervention. Konica Minolta, Cannon or Nikon could not reproduce the human, or any animals, eye or even a bulkier facsimile if they worked day and night for the next six million years. But according to evolution, random chance did!?!?#$

Also, consider this... What evolved first? The eye or the eye lid? Why evolve an eye lid without an eye? (As if *random chance* could even formulate that question, much less actually make the decision.) And if the eye evolves first then it dries up and shrivels without a lid? They would have to evolve together. At exactly the same time. In the same organism. And don't forget tear ducts. Also at exactly the same time. In the same organism. And that's exactly how it did happen. It's called creation, numbnuts.

Evolutionists can't even attempt to explain, with any level of coherency whatsoever, how the eye "evolved," or just "made itself" somehow miraculously over time. Because it absolutely could NOT have. And that IS a scientific fact. It is far too complex. Indeed its complexity is astronomical and flat out mind boggling. These video cameras can aim, focus and adjust their aperture automatically! They can see something as tiny as a human hair's diameter, and make in any given day hundreds of thousands of distinct movements. It thus affords us, up here as the conscious awareness in our brains, a continuous full 3D color picture. And then during sleep it performs any repairs and upkeep. Automatically and unconsciously. On its own.

Scientists are still searching to comprehend just how this amazing miracle of sophistication and complexity actually works. It is beyond them at this point. And keep in mind that unless all of these many complex parts of the eye are already developed, unless the eye is already fully functioning, it doesn't work. So much for "small, incremental, trial and error blind mutations" accidentally forming a complete working eye. Obviously evolution really is a joke.[101]

If anyone said that his car or his computer or his camera or his tape recorder had just evolved by random accidental chance everyone on earth would say that he was literally insane. Yet here's the kicker. The electronic and engineering equipment in your muscles, your brain, your eye, your ears, etc. are far more complicated technological wonders then these man-made

---

101   Huse, *The Collapse of Evolution,* 71-73.

artifacts. Indeed, compared to your brain, your ears and your eyes, all of our "man-made artifacts are primitive and clumsy."

> "The cells macromolecular machines contain dozens of or even hundreds of components. But unlike man made machines, which are built on assembly lines, these molecular machines assemble spontaneously from their protein and nucleic-acid components. It is as though cars could be manufactured by merely tumbling their parts onto the factory floor."[102]

How about the latest robots from Japan? They are called Dancing Japanese Robots and you can watch them on YouTube. They not only walk and dance, but can even negotiate stairs. Could the universe make these robots by random chance, starting from minerals and dirt, using small incremental accidental steps? Of course not. You'd have to be a moron to think so. Well how about a human being, that can not only make itself in the womb from just one microscopic little cell, from the information contained inside of that cell on a minute strand of DNA, unconsciously, and repair itself, completely automated, with no outside conscious intervention to help it, but it can run and walk and talk and jump and sit and negotiate a thousand obstacles hundreds of times more complicated than just a simple set of stairs or a couple of basic dance moves? Could the universe make a human being by random chance, starting from water and chemicals, using small incremental accidental steps? Of course not. You'd have to be an even bigger moron, Mr. Dawkins, to think so.

What are the odds?

"Understand, you stupid among the people: and you fools, when will you be wise? He who planted the ear, shall he not hear? He who formed the eye, shall he not see?" (Psalms 94:8-9)

Previously we have discussed the Law of Mathematical Probability and its role in disproving evolutionary theory. Here we will delve a little deeper into this fascinating subject.

If you flip a quarter, the odds of it landing on heads are fifty/fifty, or a probability of 1 in 2. If you flip that same coin a trillion times, the odds of

---

102   ("Woodson, S. A. 2005. Biophysics: assembly line inspection. *Nature* 438:566-67." As quoted on pg. 126, *The Edge of Evolution: The Search for the Limits of Darwinism,* by Dr. Michael J. Behe)

it landing *on its edge* every single time are zero. If the universe lasted for an infinite number of years, and you flipped that quarter over and over trying to get it to land on its edge a trillion straight times (I know, once would be hard enough) it would never happen. After a million years, you might even get up to 2 in a row! Those are the same odds that life, not all of life, but just one single cell, arose by random accidental chance. (Actually, your odds at the coin toss are a lot better.) If you took 1,000 jigsaw puzzles—each one having 10,000 pieces and of extreme difficulty and taking many days for a team of humans to assemble—and you took them up in a jet and dumped all of the pieces of all of those puzzles at 30,000 ft. into the jet stream over California, what are the odds of all of them landing in the same place, and assembling themselves perfectly, right next to each other, and in alphabetical order, one puzzle after another, in the parking lot of the Ravens football stadium in Baltimore, Md.? Impossible, of course, but that has a far better chance of happening accidentally by random chance than your eye happening, and just forming itself, somehow, accidentally, over time. In fact, the odds that those same Ravens will go undefeated each year and win the next 20 Super Bowls in a row by a combined score of 500 to nothing without incurring a single offensive or defensive penalty in those 380 games, are still a trillion times better odds than just one protein forming by random accidental chance outside of a living cell in some mythological, lightning-struck, primordial mud-pond.

You see, it <u>is</u> a scientific fact that macro evolutionary theory, the whole thing and just about any single part of it, is absolutely scientifically <u>impossible</u> according to the mathematical laws of probability.

As we have previously pointed out, it has been determined that statistical odds that exceed one out of 1066 can never ever happen, even if the universe lasted for an infinite number of years. The probability of <u>one</u> protein molecule forming by random chance is one in 10175. And that's just one protein…

> "The statistical improbability for the next step, the formation of a single cell from all these improbable proteins, is beyond comprehension. …It is statistically impossible for life to originate from nonlife."[103]

> "The chemical evolution of life… is ridiculously improbable.

---

103   DeRosa, *Evidence for Creation*, 53-55.

"No evolutionary theory has been able to explain why earth's atmosphere has so much oxygen. Too many substances should have absorbed oxygen on an evolving earth. Besides, if the early earth had oxygen in its atmosphere, compounds (called *amino acids*) needed for life to evolve would have been destroyed by oxidation. But if there had been no oxygen, there would have been no ozone (a form of oxygen) in the upper atmosphere. Without ozone to shield the earth, the sun's ultraviolet radiation would quickly destroy life. The only known way for both ozone and life to be here is for both to come into existence simultaneously—in other words, by creation.

"Clays and various rocks absorb nitrogen. Had millions of years passed before life evolved, the sediments that preceded life should be filled with nitrogen. Searches have never found such sediments. Basic chemistry does not support the evolution of life.

"Living matter is composed largely of proteins, which are long chains of amino acids. ...amino acids cannot link together if oxygen is present. That is, proteins could not have evolved from chance chemical reactions if the atmosphere contained oxygen. However, the chemistry of the earth's rocks, both on land and below ancient seas, shows that the earth had oxygen before the earliest fossils formed.

..."To form proteins, amino acids must also be highly concentrated in an extremely pure liquid. However, the early oceans or ponds would have been far from pure and would have diluted amino acids, so the required collisions between amino acids would rarely occur. Besides, amino acids do not naturally link up to form proteins. Instead, proteins tend to break down into amino acids. Furthermore, the proposed energy sources for forming proteins (earth's heat, electrical discharges, or solar radiation) destroy the protein products thousands of times faster than they could have formed. The many attempts to show how life might have arisen on earth have instead shown (a) the futility of that effort, (b) the immense complexity of even the simplest life, and (c) the need for a vast intelligence to precede life.

"If, despite virtually impossible odds, proteins arose by chance processes, there is not the remotest reason to believe they could ever form a membrane-encased, self-reproducing, self-repairing, metabolizing, living cell.

"There is no evidence that any stable states exist between the assumed formation of proteins and the formation of the first living cells. No scientist has ever demonstrated that this fantastic jump in complexity could have happened—even if the entire universe had been filled with proteins.

"Living cells contain thousands of different chemicals, some acidic, others basic. Many chemicals would react with others were it not for an intricate system of chemical barriers and buffers. If living things evolved, these barriers and buffers must also have evolved—but at just the right time to prevent harmful chemical reactions. How could such precise, seemingly coordinated, virtually miraculous events have happened for each of millions of species?

[Hmmm. *Kreehayshun* evolution? Wadayathink?]

"All living organisms are maintained by thousands of chemical pathways, each involving a long series of complex chemical reactions. For example, the clotting of blood, which involves 20-30 steps, is absolutely vital to healing a wound. However, clotting could be fatal if it happened inside the body. Omitting one of the many steps, inserting an unwanted step, or altering the timing of a step would probably cause death. If one thing goes wrong, all the earlier marvelous steps that worked flawlessly were in vain. Evidently, these complex pathways were created as an intricate, highly integrated system. [HELL-O!]

"The genetic information in the DNA of each human cell is roughly equivalent to a library of 4,000 books. Even if matter and life (perhaps a bacterium) somehow arose, the probability that mutations and natural selections produced this vast amount of information is essentially zero. ...To produce just the enzymes in one organism would require more than 1040,000 trials. (To begin to understand how large 1040,000 is, realize that the visible universe has fewer than 1080 atoms in it.)

"DNA cannot function without at least 75 preexisting proteins, but proteins are produced only at the direction of DNA. Because each needs the other, a satisfactory explanation for the origin of one must also explain the origin of the other. The components of these manufacturing systems must have come into existence simultaneously.

..."When a cell divides, its DNA is copied, sometimes with errors. Each animal and plant has machinery that identifies and corrects most errors; if it did not, the organism would deteriorate and become extinct. If evolution happened, which evolved first, DNA or its repair mechanism? Each requires the other."[104]

OUCH! That was a nasty blow! Evolution is reeling and down for the count. Creation has been ordered to a neutral corner. Folks, we're going to have to make the call. This fight is over.

A snowman can't even make itself. An intelligent being is required in order for it to come into existence. Someone has to roll the snow, and stack it, and stick the carrot in its face. It has to be **designed**, and then **manufactured**. No one would try to contradict this. And yet all a snowman is, is a couple of balls of snow, some buttons, a carrot and a hat. If you took trillions of video cameras and then set them up at the bottom of every slope of every snow covered mountain and hill in the world, and then waited for a snowball to roll down the hill and stop. And before it melted another ball would have to roll down and land right on top of the first without busting up either itself or the bottom snowball. Then this process would have to be repeated a third time for the head of our self-created snowman. And we're not finished. We would still need a tornado somewhere to rip a carrot out of the ground and send it through the atmosphere and fly it right into the middle of the top snowball for a nose, without damaging the three balls in the process. And then another high wind would have to snatch up a couple of acorns and send them randomly flying through the sky until they accidentally landed above the carrot in the place of two eyes. Simple enough? Trillions of times simpler than even a single celled organism. Yet with all of its simplicity, you could set up all of those cameras and you could wait an infinite number of years and you would never, ever, ever get a snowman. Never. Unless you walked around in front of one of the cameras and made the damn thing yourself. The theory of evolution is absolutely idiotic.

---

104 Brown, *In the Beginning*, 14-16.

"To claim that life evolved is to demand a miracle. The simplest conceivable form of single-celled life should have at least 600 different protein molecules. The mathematical probability that even one typical protein could form by chance arrangements of amino acid sequences is essentially zero—far less than one in 10^{450}. To appreciate the magnitude of 10^{450}, realize that the visible universe is about 10^{28} inches in diameter."

"From another perspective, suppose we packed the entire visible universe with a simple form of life, such as bacteria. Next, suppose we broke all their chemical bonds, mixed all their atoms, then let them form new links. If this were repeated a billion times a second for 20 billion years under the most favorable temperature and pressure conditions throughout the visible universe, would even one bacterium of any type reemerge? [Uh, no?] The chances are much less than one in $10^{99,999,999,873}$. Your chances of randomly drawing one preselected atom out of a universe packed with atoms are about one chance in 10^{112}—much better."[105]

Let's state the odds thing one more way. The universe has a better chance to create a car accidentally from scratch over millions of years than even one DNA molecule. It has a better chance of creating an entire small town accidentally by random chance over millions or billions of years than one single-celled amoeba. And keep in mind that these are scientific <u>facts</u>. Not theory, conjecture, speculation or wishful irrational thinking.

But what about the atheist's stubborn old argument in favor of evolution that—despite the apparent astronomical technical complexity of living cells, organs and tissues—given enough time, anything is possible? Of course, this is hogwash, for according to the mathematical laws of probability even a stupid pencil is beyond the scope of possibility for random chance to make even if given an *infinite* number of years…

"A few years ago a curious fellow decided to test the old saw that, given enough typewriters and enough time, an army of monkeys would eventually produce the works of Shakespeare. A computer with keyboard was placed in a cage containing six macaques in a British zoo and left

---

105 Brown, *In the Beginning*, 17.

for four weeks. The result? "The macaques—Elmo, Gum, Heather, Holly, Mistletoe and Rowan—produced just five pages of text between them, primarily filled with the letter S. There were greater signs of creativity towards the end, with the letters A, J, L and M making fleeting appearances." The five pages have been published under the ironic title "Notes toward the Complete Works of Shakespeare." "[106]

If it were not for their complete control of the media, the educational system and our entertainment industry, Darwinists might have already realized that they've become the laughing stock of the entire universe, and their theory would have already been exposed to mankind for the sheer lunacy that it is. And maybe then, just maybe, we could stop shoving this insanity down the throats of our naïve and unsuspecting school children in our colleges, universities, elementary and high schools. It is an insane asylum. And the inmates have taken over. (And they're called baby-mutilating, racist, black-baby-exterminating, false-accusing, lying Democrats.) (In case anyone was wondering.) ☺

Irreducible Complexity

"For you formed my inward parts; you have covered me in my mother's womb. I will praise thee; **for I am fearfully and wonderfully made**: marvelous are thy works; and that my soul knows right well." (Psalm: 139:13-14)

Irreducible Complexity is a term coined by Lehigh University biochemist Michael Behe, Ph.D. He studies the workings of the living cell on a molecular level, and is the author of two books on the subject that clearly show that the machinery inside of cells and inside of living organisms are so mind-boggling in their complexity that they are far beyond what could ever be formed accidentally or created by random mutations.

"Darwin knew that his theory of gradual evolution by natural selection carried a heavy burden: "If it could be demonstrated that any complex organ existed which could not possibly have been formed by numerous, successive,

---

106 Behe, *The Edge of Evolution,* 104-105.

slight modifications, my theory would absolutely break down."[107]

"It is safe to say that most of the scientific skepticism about Darwinism in the past century has centered on this requirement. ...What type of biological system could not be formed by "numerous, successive, slight modifications"?

"Well, for starters, a system that is irreducibly complex. By *irreducibly complex* I mean a single system composed of several well-matched, interacting parts that contribute to the basic function, wherein the removal of any one of the parts causes the system to effectively cease functioning. An irreducibly complex system cannot be produced directly (that is, by continuously improving the initial function, which continues to work by the same mechanism) by slight, successive modifications of a precursor system because any precursor to an irreducibly complex system that is missing a part is by definition nonfunctional."[108]

Dr. Behe uses the example of a Rube Goldberg contraption, which is a "deliberately <u>over engineered apparatus</u> that performs a very simple task in a very complex fashion, usually including a chain reaction."[109] Most of us have seen one of Goldberg's drawings that involve numerous successive and sometimes silly steps to put toothpaste on his brush or wipe his mouth with a napkin. The point is though, that if you were to remove just one of those many successive steps then the entire machine would be useless. The same is true with so many living systems. However, unlike a Rube Goldberg machine, an *irreducibly complex* cellular or biological system is one that is deliberately and *exquisitely* engineered to perform an extremely vital task in the most effective fashion possible, usually including many successive and highly complex steps that also can travel down multiple pathways, depending upon any number of various situations or contingencies it may encounter, in order to fulfill its function and accomplish its preprogrammed goal. But like the Goldberg contraption, if you remove just one of those steps the entire cellular or biological task fails and the organism dies. So, because of that,

---

107   Darwin, C. (1872) *Origin of Species,* 6th ed. (New York: New York University Press, 1988), 154.

108   Michael J. Behe, *Darwin's Black Box: The Biochemical Challenge to Evolution,* (New York: Free Press, 2006), 39.

109   ("Rube Goldberg machine," from Wikipedia, the free encyclopedia)

there is absolutely no conceivable process of "numerous, successive, slight modifications" that could arrive at these complex and vital cellular machines or biological systems. Either the entire machine is made at once as a fully functioning, completed system, or it is worthless, and even detrimental, to the organism. In other words, these systems could <u>never</u>, <u>ever</u>, <u>ever</u> have *evolved*, and therefore by their existence alone, Darwinism "has absolutely broken down."

> "Modern science has learned that, ultimately, life is a molecular phenomenon: All organisms are made of molecules that act as the nuts and bolts, gears and pulleys of biological systems. Certainly there are complex biological features (such as the circulation of blood) that emerge at higher levels, but the gritty details of life are the province of biomolecules. Therefore the science of biochemistry, which studies those molecules, has as its mission the exploration of the very foundation of life.

> ..."It was once expected that the basis of life would be exceedingly simple. That expectation has been smashed. Vision, motion, and other biological functions have proven to be no less sophisticated than television cameras and automobiles. [I would interject that vision, for one, is actually a billion times more "sophisticated" than any television camera. As we shall see here below...]

> ..."Life is based on *machines*--machines made of molecules! Molecular machines haul cargo from one place in the cell to another along "highways" made of other molecules, while still others act as cables, ropes, and pulleys to hold the cell in shape. Machines turn cellular switches on and off, sometimes killing the cell or causing it to grow. Solar-powered machines capture the energy of photons and store it in chemicals. Electrical machines allow current to flow through nerves. Manufacturing machines build other molecular machines, as well as themselves. Cells swim using machines, copy themselves with machinery, ingest food with machinery. In short, highly sophisticated molecular machines control every cellular process. Thus the details of life are finely calibrated, and the machinery of life enormously complex."[110]

---

110 Behe, *Darwin's Black Box*, X, 4-5.

Let's take a look at the "enormously complex machinery" of the human eye:

> "When light first strikes the retina a photon interacts with a molecule called 11-*cis*-retinal, which rearranges within picoseconds to *trans*-retinal. (A picosecond is about the time it takes light to travel the breadth of a single human hair.) The change in the shape of the retinal molecule forces a change in the shape of the protein, rhodopsin, to which the retinal is tightly bound. The protein's metamorphosis alters its behavior. Now called metarhodopsin II, the protein sticks to another protein, called transducin. Before bumping into metarhodopsin II, transducin had tightly bound a small molecule called GDP. But when transducin interacts with metarhodopsin II, the GDP falls off, and a molecule called GTP binds the transducin. (GTP is closely related to, but critically different from, GDP.)

> "GTP-transducin-metarhodopsin II now binds to a protein called phosphodiesterase, located in the inner membrane of the cell. When attached to metarhodopsin II and its entourage, the phosphodiesterase acquires the chemical ability to "cut" a molecule called cGMP (a chemical relative of both GDP and GTP). Initially, there are a lot of cGMP molecules in the cell, but the phosphodiesterase lowers its concentration, just as a pulled plug lowers the water level in a bathtub.

> "Another membrane protein that binds cGMP is called an ion channel. It acts as a gateway that regulates the number of sodium ions in the cell. Normally the ion channel allows sodium ions to flow into the cell, while a separate protein actively pumps them out again. The dual action of the ion channel and pump keeps the level of sodium ions in the cell within a narrow range. When the amount of cGMP is reduced because of cleavage by the phosphodiesterase, the ion channel closes, causing the cellular concentration of positively charged sodium ions to be reduced. This causes an imbalance of charge across the cell membrane that, finally, causes a current to be transmitted down the optic nerve to the brain. The result, when interpreted by the brain, is vision.

"If the reactions mentioned above were the only ones that operated in the cell, the supply of 11-*cis*-retinal, cGMP, and sodium ions would quickly be depleted. Something has to turn off the proteins that were turned on and restore the cell to its original state. Several mechanisms do this. First, in the dark the ion channel (in addition to sodium ions) also lets calcium ions into the cell. The calcium is pumped back out by different proteins so that a constant calcium concentration is maintained. When cGMP levels fall, shutting down the ion channel, calcium ion concentration decreases, too. The phosphodiesterase enzyme, which destroys cGMP, slows down at lower calcium concentration. Second, a protein called guanylate cyclase begins to resynthesize cGMP when calcium levels start to fall. Third, while all of this is going on, metarhodopsin II is chemically modified by an enzyme called rhodopsin kinase. The modified rhodopsin then binds to a protein known as arrestin, which prevents the rhodopsin from activating more transducin. So the cell contains mechanisms to limit the amplified signal started by a single photon.

"*Trans*-retinal eventually falls off of rhodopsin and must be reconverted to 11-*cis*-retinal and again bound by rhodopsin to get back to the starting point for another visual cycle. To accomplish this, *trans*-retinal is first chemically modified by an enzyme to *trans*-retinol--a form containing two more hydrogen atoms. A second enzyme then converts the molecule to 11-*cis*-retinol. Finally, a third enzyme removes the previously added hydrogen atoms to form 11-*cis*-retinal, a cycle is complete."[111]

And here's the kicker, "The above explanation is just a sketchy overview of the biochemistry of vision."[112]

Don't be upset if you couldn't quite understand the complexity of all of that. Indeed, <u>that's the point</u>! If you, an intelligent human being with an extremely complex computer-brain find it difficult to even comprehend the complexity of just these few initial steps in the process that leads to sight within your head, <u>how in the hell could random accidental chance ever have comprehended</u>, or "stumbled" upon, <u>it to begin with</u>? If anyone at this point

---

111   Behe, *Darwin's Black Box,* 18-21.

112   Behe, *Darwin's Black Box,* 22.

still thinks we were going a tad overboard when we said that the theory of evolution was so stupid, idiotic and ridiculous that it literally borders on the insane, then hopefully they have changed their opinion right now. For, from just those last few paragraphs alone, it is abundantly clear that the theory of evolution is not only scientifically-impossible, but it is a scientific fact that it is insane! (Richard Dawkins, are you listening? This is not a laughing matter and I am not poking fun at or mocking you now. I am telling you that you need professional help. I would suggest giving Jesus, the Great Physician, a call. He specializes in eternal diseases of the mind.)

And in case anyone might have glossed over the beginning of Dr. Behe's description, let me emphasize that he did say that all of that happens within a few **picoseconds**. And "a picosecond is about the time it takes light to travel the breadth of a single human hair." Hmm? That seems pretty damn fast. Let's see, light travels at 186,000 miles per second, and if we multiply that by 5280 ft. per mile, times 12 inches per foot, times 250 hairbreadths in an inch we get 2,946,240,000,000! That's two trillion, nine hundred and forty six billion, two hundred and forty million astronomically-complicated-sequenced chemical reactions every single second that occurs in the back of your eye in just one cell! And that is just one of a number of chemical and electrical sequences that continually occur to allow sight to happen. Multiply that by the billions of retinal cells in the back of your eyes, and the billions of human eyes and the trillions upon trillions of other animal's eyes across this planet, and this occurs every single second of every single day. The idea that this all happened by random chance is absolutely insane. It is beyond absurd, it far surpasses mere stupidity and it eclipses the idiotic and the asinine. Yea, it's insane.

Dr. Behe goes on to explain the equally enormously complex biochemical-mechanical machinery of the cilium, of blood clotting, of animal development and of intracellular transport. They rival the intricacy of sight in the human eye.

> "The function of the cilium is to be a motorized paddle. In order to achieve this function microtubules, nexin linkers, and motor proteins all have to be ordered in a precise fashion. They have to recognize each other intimately, and interact exactly. The function is not present if any of the components is missing. Furthermore, many more factors besides those listed are required to make the system useful for a living cell: the cilium has to be

positioned in the right place, oriented correctly, and turned on or off according to the needs of the cell."[113]

> "The structural elegance of systems such as the cilium, the functional sophistication of the pathways that construct them, and the total lack of serious Darwinian explanations all point insistently to the same conclusion: They are far past the edge of evolution. Such coherent, complex, cellular systems did not arise by random mutation and natural selection, any more than the Hoover Dam was built by the random accumulation of twigs, leaves, and mud."[114]

Yes but the difference is that the Hoover Dam does not have an entire false religion disguised as science built up around it that insists that Dams just *build themselves* by accident, driven of course by the unconscious struggle to create lakes and hydroelectric power.

The mechanism of blood clotting is so complicated. There are so many steps to get from a cut to a clot. There are so many options for starting the clot and stopping it at the right time. There are so many different controls and failsafe mechanisms and redundancies built into the system so a clot does not clog the artery or the veins or the capillaries and kill the organism it is supposed to heal. There are so many different pathways from which the blood clotting mechanism can go down or change course depending on each unique circumstance. To suppose that this mechanism could have made itself accidentally by random chance is absurd. It would be akin to teaching that the entire mechanized assembly line at Ford or General Motors just formed themselves, somehow, accidentally, over millions of years. No design. No intelligent foresight, direction, manufacturing of all of the many parts, and then assembly of the line into a functioning machine. Just dumb luck. (AKA Horseshit.)

> "The function of the blood-clotting system is as a strong, but transient barrier. The components of the system are ordered to that end. Fibrinogen, plasminogen, thrombin, protein C, Christmas factor, and the other components of the pathway together do something that none of the components can do alone. When vitamin K is

---

113 Behe, *Darwin's Black Box,* 204.

114 Behe, *The Edge of Evolution,* 102.

unavailable or anti-hemophilic factor is missing, the system crashes just as surely as a Rube Goldberg machine fails if a component is missing. The components cut each other in precise places, align with each other in exact ways. They act to form an elegant structure that accomplishes a specific task.

"The function of the intracellular transport system is to carry cargo from one place to another. To do this packages must be labeled, destinations recognized, and vehicles equipped. Mechanisms must be in place to leave one enclosed area of the cell and enter a different enclosed area. The failure of the system leaves a deficit of critical supplies here, a choking surplus there. Enzymes that are useful in a confined area wreak havoc in another area.

..."The designing that is currently going on in biochemistry laboratories throughout the world—the activity that is required to plan a new plasminogen that can be cleaved by thrombin, or a cow that gives growth hormone in its milk, or a bacteria that secretes human insulin—is analogous to the designing that preceded the blood-clotting system. The laboratory work of graduate students piecing together bits of genes in a deliberate effort to make something new is analogous to the work that was done to cause the first cilium."[115]

Except the design work that caused the first cilium was brought forth effortlessly by an Infinite Intelligence beyond our ability to imagine, and beyond the capability of Darwinists to demean. An Infinite Intelligence who holds in derision those who would deny Him and His Creative Power and His Divine Authorship over all of creation. "He who sits in the heavens shall laugh. The Lord shall hold them in derision." (Psalms 2:4)

"For every house is built by someone, but he who built all things is God." (Hebrews 3:4) We all understand blueprints; every construction project has them. An architect's blueprints specify everything, down to the last detail of a home addition or a skyscraper, or of any other relatively simple or highly complex construction project. Could those blueprints have ever just happened by random chance? Of course not, you'd have to be an idiot to say so. Well, every living cell has a set of construction blueprints as well.

---

115 Behe, *Darwin's Black Box,* 204-205.

The blueprints for animal bodies are inside the coded information in each animals DNA. And they are far more complicated and intricate and more ingenuously designed than any set of man-made construction plans. Could these blueprints have ever just happened by random chance? Of course not, you'd have to be a… you'd have to be taught an idiotic lie to think that as well.

> "The inadequacy of Darwinism to account for the intricacies of animal development has not been lessened by recent discoveries; it has been greatly exacerbated."[116]

> "The bottom line is that, while great progress has been made towards understanding how animals are made, and has revealed unexpected, stunning complexity, no progress at all has been made in understanding how that complexity could evolve by unintelligent processes."[117]

That's because, in a word, it can't. And this same utter incomprehensible complexity is true not only of the eye, and the blood clotting system, and the motor apparatus of cilium, and the making of an adult animal from a tiny cell, and the cargo delivery systems of the cell, but also of the immune systems of plants and animals, and the flagellum motors of bacteria, and the information coding and decoding in the DNA, and on and on and on. They are all irreducibly complex, Rube Goldberg machines to the umpteenth power. No way any single one of them, much less *all* of them, could ever have *evolved* by random accidental unintelligent chance mutations.

And if not evolution, then the only other choice is intelligent design:

> "Here, then, is the argument for design in a nutshell: (1) We infer design whenever parts appear arranged to accomplish the function. (2) The strength of the inference is quantitative and depends on the evidence; the more parts, and the more intricate and sophisticated the function, the stronger is our conclusion of design. With enough evidence, our confidence in design can approach certitude. If while crossing a heath we stumble across a watch (let alone a chronometer), no one would doubt …that the watch was designed; we would be as certain about that as

---

116   Behe, *The Edge of Evolution,* 192

117   Behe, *The Edge of Evolution,* 183.

about anything in nature. (3) Aspects of life overpower us with the appearance of design. (4) Since we have no other convincing explanation for that strong appearance of design, Darwinian pretensions notwithstanding, then we are rationally justified in concluding that parts of life were indeed purposely designed by an intelligent agent. [Ya think?]

"A crucial, often-overlooked point is that the overwhelming appearance of design strongly affects the burden of proof: in the presence of manifest design, the onus of proof is on the one who denies the plain evidence of his eyes. For example, a person who conjectured that the statues on Easter Island or the images on Mount Rushmore were actually the result of unintelligent forces would bear the substantial burden of proof the claim demanded. In those examples, the positive evidence for design would be there for all to see in the purposeful arrangement of parts to produce the images. Any putative evidence for the claim that the images were actually the result of unintelligent processes (perhaps erosion shaped by some vague, hypothesized chaotic forces) would have to clearly show that the postulated unintelligent processes could indeed do the job. In the absence of such a clear demonstration, any person would be rationally justified to prefer the design explanation.

"I think these factors account to a large degree for why, much to the consternation of Darwinian biologists, the bulk of the public rejects unintelligent processes as sufficient explanations for life. People perceive the strong appearance of design in life, are unimpressed with Darwinian arguments and examples, and will reach their own conclusions, thank you very much. Without strong, convincing evidence to show that Darwin can do the trick, the public is quite rational to embrace design. (Of course other factors besides the quality of the evidence, such as social pressure, can affect a person's judgment. In the scientific and academic communities as a whole there is strong social pressure to dismiss design explanations for life out of hand. The social situation is quite different for the general public.)"[118]

But what do evolutionists have to say about this frontal assault on their doctrine? What is the passionate defense that they have presented for all the world to hear and to debate? It doesn't exist:

> "Molecular evolution is not based on scientific authority. There is no publication in the scientific literature—in prestigious journals, specialty journals, or books—that describes how molecular evolution of any real, complex, biochemical system either did occur or even might have occurred. There are assertions that such evolution occurred, but absolutely none are supported by pertinent experiments or calculations. Since no one knows molecular evolution by direct experience, and since there is no authority on which to base claims of knowledge, it can truly be said that—like the contention that the Eagles will win the Super Bowl this year—the assertion of Darwinian molecular evolution is merely bluster.
>
> ""Publish or perish" is a proverb that academicians take seriously. If you do not publish your work for the rest of the community to evaluate, then you have no business in academia (and if you don't already have tenure, you will be banished). But the saying can be applied to theories as well. If the theory claims to be able to explain some phenomenon but does not generate even an attempt at an explanation, then it should be banished. Despite comparing sequences and mathematical modeling, molecular evolution has never addressed the question of how complex structures came to be. In effect, the theory of Darwinian molecular evolution has not published, and so it should perish."[119]

> "The impotence of Darwinian theory in accounting for the molecular basis of life is evident not only from the analyses in this book, but also from the complete absence in the professional scientific literature of any detailed models by which complex biochemical systems could have

---

118 Behe, *Darwin's Black Box*, 265-266, 309.

119 Behe, *Darwin's Black Box*, 185-186.

been produced… In the face of the enormous complexity that modern biochemistry has uncovered in the cell, the scientific community is paralyzed. No one at Harvard University, no one at the National Institutes of Health, no member of the National Academy of Sciences, no Nobel Prize winner—no one at all can give a detailed account of how the cilium, or vision, or blood clotting, or any complex biochemical process might have developed in a Darwinian fashion. But here we are. Plants and animals are here. The complex systems are here. All these things got here somehow: if not in a Darwinian fashion, then how?"[120]

Hmm, I'm going to take a wild stab in the dark at that one but, uh, anyone ever heard of Genesis? The Bible? Mr. Hitchens, ever heard of God??

"If the great majority of cellular protein-protein interactions are beyond the edge of evolution [beyond the ability of random chance to *ever* produce], it is reasonable to view the entire cell itself as a nonrandom, integrated whole—like a well-planned factory, as National Academy of Sciences president Bruce Alberts suggested. … Nonrandomness isn't a rare property of just a handful of extra-complex features of the cell. Rather **it encompasses the cellular foundation of life as a whole**."[121] (Emphasis mine.)

What that means is that if it is a scientific <u>fact</u> that random chance could never ever have produced the complexity of life's cellular systems (and it is), then evolution is impossible and the only other plausible scientific theory that remains for the origin and diversity of life is that it was deliberately and purposefully <u>designed</u> and then <u>manufactured</u> by an intelligence and a power far beyond our ability to comprehend. All we can do is to stand back and look on in absolute awe and wonder. An awe and a wonder and a respect that is being stolen from our school children every day across this nation by the shoveling down their throats of the bald-faced, idiotic, asinine, scientifically-impossible, poisonous, noxious, satanically-inspired, rancid lie that is the theory of evolution! Congratulations, America! My, how proud you must be… "My boy Charlie is an honor student at Leftos Indoctrinato

---

120   Behe, *Darwin's Black Box*, 187.

121   Behe, *The Edge of Evolution*, 146.

Demonrato Junior High, and he got straight A's in evolutionary biology." And if you had just a tad more sense and a little more actual education then you might have been a little more proud if he had told his teacher she was wrong, that she was teaching a lie as if it were true, and had gotten straight F's instead. (Fascists are highly intolerant of dissent.)

Well there you have it. I think that should be more than sufficient for anyone but the saddest of religious fanatics. For that is what in truth the theory of evolution is. It is a religion and not science. And a false and absurd religion at that. And keep in mind that we have only scratched the surface of the tip of the iceberg in our discussion of the scientific evidence here. (See the list of recommended books at the end of this chapter, many of which we have quoted from.) For there is in truth a crushing avalanche of scientific facts and evidence that obliterates the *theory* of evolution. And there are no scientific facts or evidence that can be used to credibly support this theory. It is an unscientific and blind religious faith in impossible happenstance, in order to afford its proselytes the dubious distinction of denying their own Creator.

"But ask now the beasts, and they shall teach you; and the fowls of the air, and they shall tell you. Or speak to the earth, and it shall teach you: and the fishes of the sea shall declare unto you. Who knows not in all of these that the hand of the Lord has wrought this?" (Job 12:7-9) If the beasts of the field and the fishes of the sea can figure out who wrought all of this, you'd of thought it would have at least by now been within the grasp of the intellectual elites in our universities and in our media. You'd of thought wrong.

The Big Bang, All Blown Up

"God, who at many times and in various ways spoke in time past unto the fathers by the prophets, has in these last days spoken unto us by his Son, whom he has appointed heir of all things, <u>through whom also He</u> [and not some stupid explosion] **<u>made</u>** <u>the universe</u>. Who being the brightness of his glory and the exact image of his being, <u>sustaining all things by his powerful word</u>, when he had by himself purged our sins, sat down on the right hand of the Majesty on high." (Hebrews 1:1-3)

Well, I think poor evolution has had about all it can take. So let's give it a little space to expire peacefully as we shift our attention now to its companion theory and the other half of Satan's genesis, the big bang. This theory states that the entire universe was created by an enormous explosion billions of years ago and that in the proceeding eons of time everything you see today in the night sky, including our planet, just made itself; just *evolved* somehow accidentally and miraculously by random chance. Of course this is almost as idiotic as the notion that life just made itself, and equally as scientifically-impossible, as we shall see…

Again we start with the Laws of science:

> "The Second Law of Thermodynamics… tends to bring a system to disorder. …When one observes the universe, the second law is apparent everywhere. The sun is wearing down slowly; stars are burning out and even exploding. … Big Bang theory contradicts the Second Law because it requires particles to organize and cohere on a cosmic scale. There is no scientific evidence for this claim. …What we see in the universe is directly opposite to the expectation of evolutionary cosmologists. We observe a decaying universe whose order of complexity is in decline. Evolution cosmology directly defies this great law of science."[122]

Evolution cosmology also defies all of the facts and evidence, and common sense as well…

Faux science is enamored with the theory of the big bang, popularized by the British scientist, Stephen Hawking. Many people have heard of his bestseller, *A Brief History of Time*, "a book that was widely considered fascinating by those who did not read it, and incomprehensible by those who did."[123] It is "incomprehensible" because it champions a theory that is clearly scientifically impossible. There is no such thing as an explosion ever creating anything of any value or order (See chapter 5 "God" in the larger book). All an explosion can do is to make a big mess—to tear apart, to break down and to destroy—and this universe is about the furthest thing from a big mess you will ever find. Indeed, the universe displays order to an almost unimaginable level. When you learn of all of the laws that govern motion, and gravity, and light, and sound, and that control the behavior of atoms and molecules and their electrons, protons and neutrons; the order and engineering complexity of this universe is *astronomical* and beyond our comprehension. Also, according to the very laws of science itself, in particular the Second Law of Thermodynamics and the Laws of Mathematical

---

122   DeRosa, *Evidence for Creation*, 30.

123   Berlinski, *The Devil's Delusion*, 98.

Probability, we know for a scientific fact that it is an absolute impossibility for this universe, this solar system, our Earth-Moon system, and the subsequent capacity of our planet to be inhabitable, to have been created by an ancient, cosmic explosion.

Again (as we reported in chapter 5 in "All the Secrets") Cambridge University physicist Brandon Carter PhD calculated "that if gravity had been stronger or weaker by one part in 10^{40}, then life-sustaining stars like the sun could not exist. This would most likely make life impossible."[124] (This would *definitely* make life impossible.) 10^{40} is one with 40 zeroes behind it. A trillion is one with only 12 zeroes behind it. 10^{40} is a trillion times a trillion times a trillion times ten thousand. The chances of winning the lottery are around a million to one. The chances of the force of gravity accidentally being <u>exactly</u> the force needed to render the universe inhabitable is trillions and trillions and billions of times more remote than the chances of you winning the lottery. Try to shoot an arrow across the entire known universe with your eyes closed and hit a target in the <u>exact</u> center without being off by even one trillionth of an inch. And then try to do it a billion times in a row. It's impossible. So is the accidental formation of this universe. The force of gravity was specifically created to be <u>exactly</u> the amount needed by a degree of variance of less than one over 10^{40}. And this is only one of a number of examples of the mind-numbing exactness of the laws and constants of this universe. (In addition to the gravitational force mentioned above that holds matter together there is also the three forces—the electromagnetic force and the strong and the weak nuclear forces—that hold the structure of the atom together which are also <u>exactly</u> calibrated so the heavier elements can form and we can have a universe of matter that is able to form stars and planets and to support life.) "In light of these and other examples, Collins remarks that 'Almost everything about the basic structure of the universe ... is balanced on a razor's edge for life to occur.'"[125] God, 10^{40}. Big bang theory, zip. Game, set, match. Without stars and without orbiting planets, there could be no earth and there could be no life.

"God is a mathematician of a very high order and he used advanced mathematics in constructing the universe."[126] Again, what an understatement.

---

124 (*International Journal of Systematic and Evolutionary Microbiology*, IJSEM, Collins 1999, 49)

125 (IJSEM, Collins 1999, 48)

126 (Paul A. M. Dirac)

> "There is no empirical evidence to support the star formation theory proposed by evolutionary cosmologists. [It is no more than wishful thinking, using one's fertile imagination to try and make sense of a scientifically-impossible theory.] ... The explanation offered is that as cooling occurs, particles slow down and clump together. The problem is, however, that the celestial objects are moving at relatively high speeds away from each other. No star or galaxy has ever been seen to form in space from star gas. As Harvard astrophysicist, Abraham Loeb stated, "The truth is that we don't understand star formation at a fundamental level." "[127]

The truth is they never will "understand star formation at a fundamental level" until they understand that someone of infinite intelligence and power *formed* the damn things to begin with.

"The heavens declare the glory of God; and the firmament shows his handiwork." (Psalms 19:1)

There is no compelling or even credible scientific evidence for, nor is there even an intelligent idea of, how the universe could have just formed itself. Again, both our solar system and the universe are <u>highly ordered</u>. And order <u>cannot</u> come out of disorder by an explosion, or accidentally, or by random chance, or by any combination thereof. Thus speaks The Second Law of Thermodynamics, The Laws of Mathematical Probability, logic, common sense, and as we shall see, all of the facts and evidence as well.

Scott Huse in his book The Collapse of Evolution lists a number of features of our earth sun system that together defy any random, accidental chance explanation. But instead point to the fact that it had to be specifically and meticulously designed...

-The earth's almost circular orbit as well as our atmospheric CO_2 and water vapor, keeps us from extreme, life-killing temperature fluctuations.

-Ours is the only planet in the solar system that gets the perfect heat amount for sustaining life. The others are either too close and therefore too hot, or too far and therefore too cold.

-Also, if the rate of our daily rotation were to be changed by even a factor of 1/10th all vegetable life would either freeze at night or fry by day.

---

127   DeRosa, *Evidence for Creation*, 30.

-Change the distance of the moon from the earth by just 1/5th and instead of vital and innocent tides, the continents two times a day would be underwater.

-Change the mass and size of the earth just slightly and the balance between the pressure of the atmosphere and gravity would make life difficult at best.

-Water is life, without it life could not exist, and our earth is blessed with a more than generous supply. Water doesn't occur in the universe accidentally by random forces of nature. It has to be placed there.

-Life is protected from the sun's deadly radiation by our magnetic field.

-The earth's atmosphere, which also protects life from roughly 20 million meteors travelling at 30 miles a second each day, consists of an essential ratio of 20% oxygen and 78% nitrogen which is critical to all life.

-Our ozone layer is a defense against the sun's ultraviolet radiation and therefore all of life is not destroyed.

-And finally the tilt of our axis at 23 degrees along with our slightly elliptical orbit allows for our seasons. Without which our agricultural supply of food would be dramatically altered.

These are just a few of the examples we could list that point undeniably to an intelligent creation of this earth-sun system for the existence and sustaining of the incredible and astronomical variety of life forms. The faux biology scientist's blind belief that this support system could have just created itself, somehow, accidentally, by random chance is totally mindless.[128]

But don't tell Mr. Evolutionist that. It might hurt his feelings. And protecting his feelings is so much more important than stopping him from continuing to lie to our naïve, gullible and unsuspecting school children in our colleges, universities, elementary and high schools, about the existence of an intelligent Creator called God. After all, not *that* many of our kids commit suicide because of his lie. And not all *that* many of them end up hooked on drugs for the rest of their lives or fall into debased sexual immorality because of the purposelessness that their lie has instilled in them. So go ahead, America, continue blissfully in your stupidity and ignorance and just let this lie be taught unchallenged and unabated. Oh, gosh! I'm sorry, America, now I've gone and hurt your feelings.

---

128   Huse, *The Collapse of Evolution,* 55-58.

"The evolutionary "Big Bang" theory proposes that dust particles in space coalesced to form galaxies, stars, planets and other celestial objects, and that 4.5 billions of years ago, the beginnings of our sun (a "proto-sun") formed out of a large gas cloud. [All of which is physically impossible.] The leftover gas from the cloud flattened into a disc [miraculously] [why the gas chose the disc over a cube, a pyramid, a pretzel or a Botticelli, remains a mystery], and then little particles coalesced, got bigger and bigger and developed into planet-like clumps that eventually became the planets of our solar system. Unfortunately for Big Bang proponents, this "common origin" theory of our solar system directly conflicts with known facts about the planets, including their vastly different compositions, rotations, distances between them, densities, magnetic fields, etc. The situation is summarized by physicist Lambert Dolphin, who states, "Theories for the origin of our solar system have come and gone, yet even today no satisfactory model exists that both explains all the facts and is consistent with the known laws of physics." In fact, physical laws inform us that far more than gravity is required to form a star, planet or any other heavenly body out of particles in a giant gas cloud. [You also need pixie dust.] Simple physics tells us that (kinetic) energy tends to disperse particles in a gas cloud, or cause them to move away from each other, not bring them together. In other words, gas clouds naturally expand rather than condense. It is actually preposterous—and defies the laws of science— to believe that without an overwhelming outside force, gas can contract to the point where nuclear fusion can take place, as in our sun."[129]

"Many undisputed observations contradict current theories on how the solar system evolved."[130] And remember, in science as in life, observations trump theories all day long. If a theory is not backed up by our observations, then that theory must go. According to the theory of the big explosion and the subsequent "evolution" of the stars and galaxies of this universe, and of

---

129 (From Creation Studies Institute circular, March 2011, Tom DeRosa, Executive Director and Founder)

130 Brown, *In the Beginning,* 25.

our own solar system, we should expect that certain things would be true. But they are not…

> "All planets should spin in the same direction, but Venus, Uranus, and Pluto rotate backwards. …Each of the almost 200 known moons in the solar system should orbit its planet in the same direction, but more than 30 have backward orbits. Furthermore, Jupiter, Saturn, Uranus, and Neptune have moons orbiting in both directions. … The orbit of each of these moons should lie very near the equatorial plane of the planet it orbits, but many, including the Earth's moon, are in highly inclined orbits. …The orbital planes of the planets should lie in the equatorial plane of the Sun. Instead, the orbital planes of the planets typically deviate from the Sun's equatorial plane by 7 degrees, a significant amount. …The Sun should have about 700 times more angular momentum than all the planets combined. Instead, the planets have 50 times more angular momentum than the Sun."[131]

There is no known way for our earth to have acquired all of its water by natural, evolutionary means. Every explanation that the evolutionary cosmologists try to come up with turns out to be absurd. Comets couldn't have "brought" it here, neither could meteorites, neither could any other fantasized water transport system. It had to be placed here in the beginning, plain and simple.

> "If the Earth and solar system evolved from a swirling cloud of dust and gas, almost no water would reside near Earth's present orbit. Any water (liquid or ice) that close to the Sun would vaporize and be blown by solar wind to the outer reaches of the solar system, as we see happening with water vapor in the tails of comets."[132]

Textbooks teach our children that the early earth was molten and bombarded by meteorites. But that is false as well:

---

131 Brown, *In the Beginning*, 25.

132 Brown, *In the Beginning*, 26.

"Had Earth ever been molten, dense, nonreactive chemical elements such as gold would have sunk to Earth's core. Gold is 70% denser than lead, yet it is found at the Earth's surface."[133]

Also, radioactive dating of certain zircon minerals, that had to have formed on a <u>cold</u> earth, contradicts a molten "early" Earth.

"Evolutionists claim that the solar system condensed out of a vast cloud of swirling dust about 4,600,000,000 years ago." This however is scientifically and astronomically impossible:

"Contrary to popular opinion, planets should not form from just the mutual gravitational attraction of particles orbiting the Sun. Orbiting particles are much more likely to be scattered or expelled by their gravitational attraction than they are to be permanently pulled together. Experiments have shown that colliding particles almost always fragment rather than stick together. (Similar difficulties exist in trying to form a moon from particles orbiting the planet.)

"Despite these problems, let us assume that pebble-size to moon-size particles somehow evolved. "Growing a planet" by many small collisions will produce an almost *nonspinning* planet, because spins imparted by impacts will be largely self-canceling."

"The growth of a large, gaseous planet (such as Jupiter, Saturn, Uranus, or Neptune) far from the central star is especially difficult for evolutionist astronomers to explain…

…"Based on demonstrable science, gaseous planets and the rest of the solar system did not evolve."[134]

"Thou burning sun with golden beam, Thou silver moon with softer gleam, Oh praise Him! Oh praise Him!" (Francis of Assisi, circa 1225)

And then there's Saturn…

---

133   Brown, *In the Beginning*, 26.

134   Brown, *In the Beginning*, 27-28.

> "Planetary rings... have nothing to do with a planet's origin. ...Supposedly after planets formed from a swirling dust cloud, rings remained... [But that is just not so.] Rings form when material is expelled from a moon by a volcano, a geyser, or the impact of a comet or meteorite. Debris that escapes a moon because of its weak gravity and a giant planets gigantic gravity then orbits that planet as a ring. If these rings were not periodically replenished, they would be disbursed in less than 10,000 years. Because a planet's gravity pulls escaped particles away from its moons, particles orbiting the planet could never form moons--as evolutionists assert."[135]

And the score at halftime is: Planetary rings, 10,000. Big bang, nothing. Let's see what strategy big bang has up his sleeve for the second half. Ah! He and coach Hawking have gone to the chalkboard, and are drawing feverishly...

> "Evolutionary theories for the origin of the Moon are highly speculative and completely inadequate. The Moon could not have spun off from Earth, because its orbital plane is too highly inclined. Nor could it have formed from the same material as Earth, because the relative abundances of its elements are too dissimilar from those of Earth. The Moon's nearly circular orbit is also strong evidence that it was never torn from nor captured by Earth. ...[Nor was the Moon torn off by] a Mars-size impactor.
>
> ..."These explanations have many other problems. Understanding them caused one expert to joke, "The best explanation is observational error--the Moon does not exist." Similar difficulties exist for evolutionary explanations of the other (almost 200) known moons in the solar system.
>
> "But the Moon does exist. If it was not pulled or splashed from Earth, was not built up from smaller particles near its present orbit, and was not captured from outside its present orbit, only one hypothesis remains: the Moon was created in its present orbit."[136]

---

135   Brown, *In the Beginning,* 27.

136   Brown, *In the Beginning,* 27-28.

"He. Could. Go. All. The. Way." Yes! It's another touchdown. In the "Origin of the Moon" game, Creation has scored dozens of touchdowns, extra points and field goals, and evolution has yet to get a first down. But that's OK! Do not let your hearts be troubled! They will still lie to our school children and tell them evolution has won. Ooh, I caint wait 'till the next time I can go out and vote Democrat! How about you? (The theory of evolution is *the* foundation of the godless, baby-mutilating, lying, racist, black baby exterminating Democrat-Left, as we will see in a later section of this chapter.)[137]

> "The Big Bang theory [is] now known to be seriously flawed...
>
> "The redshift of starlight is usually interpreted as a Doppler effect; that is, stars and galaxies are moving away from Earth, stretching out (or reddening) the wavelengths of light they admit. Space itself supposedly expands— so the total potential energy of stars, galaxies, and other matter increases today with no corresponding loss of energy elsewhere. Thus, the big bang violates the law of conservation of energy [or The *First* Law of Thermodynamics], probably the most important of all physical laws.
>
> ..."Many objects with high redshifts seem connected, or associated, with objects having low redshifts."[138]

This, and a number of other observational evidences, is **in**consistent with the Doppler effect and with the big bang theory in general. Evolutionists also teach that the *cosmic microwave background* (CMB) *radiation* is evidence of the big bang theory. That is false...

> "Because the CMB is so uniform, many thought it came from evenly spread matter soon after a big bang. But such uniformly distributed matter would hardly gravitate in any direction; even after tens of billions years, galaxies and much larger structures would not evolve. In other words, the big bang did not produce the CMB.

---

137 (And you can also see Chapter 11, "Baby Mutilation," in the larger work on the truth)

138 Brown, *In the Beginning,* 30.

"Contrary to what is commonly thought, the big bang theory does not explain the amount of helium in the universe; the theory was adjusted to fit the amount of helium. Ironically, the lack of helium in certain types of stars (B type stars) and the presence of beryllium and boron in "older" stars contradicts the big bang theory.

"A big bang would produce only hydrogen, helium, and lithium, so the first generation of stars to somehow form after a big bang should consist only of those elements. Some of these stars should still exist, but despite extensive searches, none has been found.

…"If the big bang occurred, we should not see massive galaxies at such great distances, but such galaxies are seen. …A big bang should not produce highly concentrated or rotating bodies. Galaxies are examples of both. Nor should a big bang produce tightly clustered galaxies. Also, a large volume of the universe should not be—but evidently is— moving sideways, almost perpendicular to the direction of apparent expansion."[139]

The evidence just keeps pouring in, but as fast as it does, big bang cosmologists just keep ignoring it, or dismissing it out of hand. (How they get away with doing that to the observable evidence, while still calling themselves *scientists,* we've yet to figure out.) (Let's ask Jon Stewart. He's so smart. Maybe he knows.) "Hear no evil. See no evil. Hear no evidence. See no facts. Monkey see, monkey do. Evolution is a fact. Evolution is a fact. Evolution is a fact…" Remember, people believe whatever they are taught and, *if you shout a lie loud enough and long enough* and academically blackball anyone who dissents,[140] your lie will stick.

"Evolutionists claim that stars form from swirling clouds of dust and gas. For this to happen, vast amounts of energy, angular momentum, and residual magnetism must be removed from each cloud. This is not observed

---

139   Brown, *In the Beginning,* 30.

140   (See *Expelled:* **No Intelligence Allowed,** a documentary narrated by Ben Stein that exposes the fascism of today's evolutionists who control academia, and who not only don't tolerate dissent, but are openly hostile to it and punish anyone foolish enough to cross their doctrinal beliefs. Of course, the fact that doing so is the antithesis of scientific inquiry doesn't appear to phase them in the least.)

today, and astronomers and physicists have been unable to explain, in an experientially verifiable way, how it all could happen.

…"If stars evolve, star births should about equal star deaths. …We have seen hundreds of stars die, but we have never seen a star born."

Also, stars could not have "evolved in globular clusters, where up to a million stars occupy a relatively small volume of space. …Wind and radiation pressure from the first star in the cluster to evolve would have blown away most of the gas needed to form the other stars in the cluster. In other words, if stars evolved, we should not see globular clusters, yet our galaxy has about 200 globular clusters. To pack so many stars that tightly together requires that they all came into existence at about the same time.

"Poor logic is involved in arguing for stellar evolution, which is assumed in estimating the ages of stars. These ages are then used to establish a framework for stellar evolution. This is circular reasoning.

"In summary, there is no evidence that stars evolve, there is much evidence that stars did not evolve, and there are no experimentally verifiable explanations for how they could evolve and seemingly defy the laws of physics."[141]

So the big bang theory is not only absurd according to common sense and our everyday experience, it is scientifically impossible according to the very laws of science itself—the first and second laws of thermodynamics and the mathematical laws of probability that tell us that explosions can't create anything of order. So an explosion could not and did not create the universe. But the big bang theory is also false according to all of our *observations* of the universe itself. And the efforts of evolutionary cosmologists to ignore this evidence, and their feeble attempts to try to *fix* the many problems these observations have created for them (only a few of which we have listed here) have been tedious and at times absurd. It is also professionally incriminating as to their close-mindedness at even entertaining the thought that there might exist another theory of the origin of the universe that is in total agreement with both the laws of science and with all of the observational data: Creation. But entertaining anything other than their accepted dogma is beyond the

---

141   Brown, I*n the Beginning,* 32.

scope of religious zealots (in this case, devotees to the religion of *Naturalism*[142]) so they nevertheless doggedly persist in their vain and useless attempts at supporting the insupportable. The big bang theory is a big bust. Why then is it still taught as *fact* to our unsuspecting and gullible school children in our colleges, universities, elementary and high schools?

> "The hard work of many scientists across many scientific disciplines in the past century unexpectedly demonstrated that both the universe at large and the earth in particular were designed for life. The heavens and earth—and life itself—alike are fine-tuned."[143]

"And to make all men see what the fellowship of the mystery is, which from the beginning of the world has been hid in God, <u>who created all things through Jesus Christ</u>." (Ephesians 3:9)

<u>- A Bounded Universe</u> Another somewhat complicated but nevertheless elucidating fact about this subject of cosmology that has been completely ignored by our disseminators of false information is that a fundamental mathematical error entered into the equations of evolutionary cosmologists from the outset and thus falsified their conclusions. Conclusions that led them to the idea of that *big, primordial explosion.* This error stems from their erroneous assumption that we live in what has been called an *unbounded* universe.

> "Most non-experts (in fact, most scientists not trained in cosmology) are unaware that the universe assumed by the Big Bang theorist has no boundaries, no edge and no center.
>
> ..."Why do Big-Bang cosmologist use as their starting point the assumption (which seems quite contrary to common sense) that the universe has no boundary? Is there some good scientific reason, or is it perhaps demanded or even suggested by well-established, experimentally-backed theory, like general relativity?

---

142   (We mentioned this religion earlier in the beginning of chapter 1 "Truth," in the larger work, and will discuss it further at the end of this chapter)

143   Behe, *The Edge of Evolution*, 218-219.

The answer is no. It is an *arbitrary assumption,* called the "cosmological principle," or more recently, the "Copernican principle." This assumes that (whether the universe is finite... or infinite) there is no edge and no center. On a large enough scale matter is evenly distributed around us."[144]

And the reason they have assumed that is because what we *do* observe from our vantage point here on earth is that all of the matter in the universe (stars and galaxies) does appear to be evenly distributed all around us. Therefore, rather than entertaining the thought that the even distribution of matter around us belies the fact that our earth is in a special and unique position at the center of God's universe (which of course it is), big bang cosmologists instead jumped to the more secularly acceptable assumption (God being such a hard pill for them to swallow) that there is no edge and no center. Because if there *were* an edge and a center then...

...."why don't we see more matter and more galaxies on one side of us than the other? This would be easy to explain if we were in a special place close to the center. ...[However] Such a "special arrangement" is exceedingly improbable on a chance basis. It therefore strongly smacks of purpose, and [here's the kicker] is **thus unpalatable to most theorists today, who prefer to believe in a universe ruled by randomness**. [Emphasis mine] [Which is far better than a universe ruled by God because, you see, you don't have to answer to randomness.] ...Of course, this idea of randomness is the essence of Darwinism... So it is simply assumed that there *is* no center, and no boundary. In this assumption, every part of the universe will appear to have matter evenly distributed around it as well.

...."It may not be unfair to suggest another possible reason for the near-universal acceptance of this assumption. To allow the possibility of anything "outside" the universe (perhaps God?) makes it harder to hold the position that the universe is "all there is" (the popular position of philosophical materialism)."[145] [And a popular position

---

144 D. Russell Humphreys, Ph.D., *Starlight and Time: Solving the Puzzle of Distant Starlight in a Young Universe* (Green Forest, AR: Master Books, 2006), 14, 18-19.

145 Humphreys, *Starlight and Time,* 19, 87.

that was destroyed in Chapter 3 "The Spiritual Universe"
in the larger book, "All the Secrets."]

So it is easy to see what has happened here. Because our cosmological scientists are in love with Mr. Random Chance, and worship him as the creator of the universe, and of course of life here on earth, they refuse to acknowledge that this universe was created with a purpose and that the creation of life, and especially man, is at the forefront of this purpose. And that, therefore, of course this extraordinarily unique water planet is at the center of this universe! And therefore of course we would see the same amount of stars and galaxies, and matter, all around us.

Conversely, if one starts out with these correct assumptions then, mathematically, we get an entirely different cosmology. And one that also neatly coincides, rather than blatantly conflicts, with all of the observed data:

> "What if we begin our calculations with the opposite assumption, equally scientifically valid, namely that matter in the universe *has* a center and an edge (is bounded)? This makes more common sense and is also Scripturally far more appropriate. When we feed in this, plus the same observations, into general relativity, quite a different cosmology falls out."[146]

And that cosmology is one where God "stretched out the heavens like a curtain." (Isaiah 40:22) It never *exploded*. He created matter and then formed the stars and the galaxies and stretched them out to their present positions. This is one of the conclusions that are reached when the assumption of a *bounded* universe with earth at its *center* is plugged into the general theory of relativity. It is also a historical fact of creation according to God's Word which we have already determined to be scientifically, and historically, accurate. (See chapter 6, "The Word of God," again, in the larger work.) Besides the verse in Isaiah quoted above, there are numerous others that reveal this. Here are a few:

"Who alone <u>spreads out the heavens</u>, and treads upon the waves of the sea." (Job 9:8)

"Who covers yourself with light as with a garment, Who <u>stretches out the heavens like a curtain</u>." (Psalms 104:2)

---

146 Humphreys, *Starlight and Time*, 21.

"He has made the earth by his power, he has established the world by his wisdom, and has <u>stretched out the heavens at his discretion</u>." (Jeremiah 10:12)

"Thus says the LORD, <u>who stretches out the heavens</u>, and lays the foundation of the earth, and forms the spirit of man within him." (Zechariah 12:1)

"Indeed My hand has laid the foundation of the earth, and <u>My right hand has stretched out the heavens</u>." (Isaiah 48:13)

"I have made the earth, and created man upon it: <u>I, even my hands, have stretched out the heavens</u>, and all their host have I commanded." (Isaiah 45:12)

"Thus says God the LORD, he <u>who created the heavens and stretched them out</u>, he who spread forth the earth and that which comes from it, he who gives breath to the people upon it, and spirit to those who walk therein." (Isaiah 42:5)

This "stretching out" or expansion also caused the redshifts of distant stars and galaxies.

> "A... misconception is that the red shifts of the galaxies are Doppler shifts, i.e. caused by the velocity of the galaxies away from us at the time the light starts its journey toward us. But [even] one undergraduate textbook... and many graduate textbooks... make it clear that the red shifts are an *expansion* effect. As space is stretched out, the lengths of all electromagnetic waves passing through the space are similarly stretched out. Consequently, the *speed* of recession doesn't matter, only the *amount* of expansion that takes place as the light travels to us, whether the expansion is fast or slow."[147]

It is also the reason, along with gravitational time dilation, why we can see the light from those distant galaxies (some as far as 12 billion light years away) even though our earth and our universe are very young, less than 10,000 years old. (See upcoming sections, "The Young Earth" and "Starlight and Time.")

So, if you feed the <u>ideological assumption</u> of an **un**bounded universe into Einstein's theory of general relativity then mathematically you will come out with the absurd theory of the big bang—that an explosion made

---

147   Humphreys, *Starlight and Time*, 98.

the universe—which as we have seen is also scientifically impossible and completely unsatisfactory as a creator of all that we see now in this highly ordered and complex universe, solar system, earth-moon system, and living planet.

Conversely, the mathematical result of feeding into the general theory of relativity the accurate assumption that the universe is bounded, has a center and an edge, and that the earth is at this center, is that "the cosmos has been expanded" (and *not* exploded) and that the universe is very young.[148] And, coincidentally, that is exactly what the Bible says.

Keep in mind also that the theory of the big bang was not hypothesized in a vacuum, or in the pure unbiased search for scientific truth. It was formulated under the weight of the extreme bias of 1) assuming evolution was true, and 2) assuming that the universe had to have formed itself as well and therefore *had to be* billions of years old in order to have had the time to do that. We know now that the first assumption is absolutely false, and the second assumption has already been dealt a fatal blow and will soon expire as we continue through the remainder of this book.

(And if after all this, anyone still believes in the "Big Bang," and that this mythological "primordial" explosion created the universe, accidentally, by random chance, then may I suggest you get some dynamite, go into the woods, and build yourself a housing development…)

The Young Earth

"He stretches out the north over the empty place, and hangs the earth upon nothing."(Job 26:7) (…And He did it about 6 thousand years ago, and not hundreds of millions or billions.)

We have already introduced a number of facts and evidences throughout this book that contradict the established conclusions of the main-stream of the scientific community (at least that part of the scientific community that maintains near-totalitarian and fascist control of our media, our entertainment industry and our educational system) that the earth and this universe are "billions of years old." But in this section we will introduce enough further evidence that will establish that the age of the earth, the solar system and the universe is less than ten thousand years old. Again, the evil one's origins account poisons our educational system, and those who pass through it have become contaminated with the Darwinist drivel they

---

148 Humphreys, *Starlight and Time,* 79-80, 99-100.

were taught. Indeed, the unscientific but evolutionary compatible lie that the universe is billions of years old has been so universally shoveled into the minds of America, and from every conceivable source, in order to firmly establish it as another so-called scientific fact, that it is of great importance to provide as much information here as possible so we can shovel it all back out. Here we go…

The consensus among scientists, up until the 19th century, was that the earth was only thousands of years old, and that the evidence for the catastrophism of the Genesis Flood was everywhere and undeniable. But then along came James Hutton followed by Sir Charles Lyell with the concept called *uniformitarianism*[149] that said, erroneously, that all the earth's features can be explained by the same slow natural progressions observable today. The phrase "the present is the key to the past" became the rallying cry of this dogma and is still taught today. Today's historical geology is dominated by this concept which has caused the earth to "age" in just over a century from thousands of years to the astronomically inflated, mythological and blindly believed, 5 billion years.[150]

> "Most Scientific Dating Techniques Indicate That the Earth, Solar System, and Universe Are Young. For the last 150 years, the age of the Earth, as assumed by evolutionists, has been doubling at roughly a rate of once every 15 years. In fact, since 1900 this age has multiplied by a factor of 100!"[151]

That is because their theory collapses without hundreds of millions and billions of years for evolution to work its *miracle*. (But even that is moot because we now know that evolution collapses no matter how many eons you give it.)

> "Our media and textbooks have implied for over a century that these almost unimaginable ages are correct. Rarely do people examine the shaky assumptions and growing body of contrary evidence. Therefore, most people today almost instinctively believe that the Earth and universe are billions of years old. Sometimes, these

---

149   (similar to *antidisestablishmentarianism* but much shorter)

150   Huse, *The Collapse of Evolution*, 7.

151   Brown, *In the Beginning*, 37.

people are disturbed, at least initially, when they see the evidence."[152]

"And I will send them strong delusion, that they should believe a lie." The Creator is basically saying, *you want lies, you got lies. You love them more than me, then help yourself. Go back for seconds. You want to ignore the evidence and believe a fairy tale, then go ahead. I won't stand in your way. I gave you free will, and you are free to choose the bald-faced lies of Satan over my scientific truth.* (Where does scientific truth come from anyway? Scientists? Guess again.) You get what you ask for. Which is why it is so important at some time in your life, no matter what you were taught, to ask for the truth. So you can receive.

Let's look at some facts:

> "Humanlike footprints, supposedly 150--600 million years old, have been found in rock formations in Utah, Kentucky, Missouri, and …Pennsylvania."[153]

Of course these supposed ages for the rock predate man by hundreds of millions of years, according to evolutionary time lines. Obviously something is very wrong here. ("We have *another* problem, Houston.")

The magnetic field of the earth has been continuously weakening. The decay rate matches a 1400 year half-life. What this means is that if we go back just 10,000 years our magnetic field would be the equal of a magnetic star! And this is highly unlikely/nearly impossible. Therefore a 10,000 year limit for the earth's age would seem to be accurate according to our magnetic field's decay rate. And if evolutionists want to reject this logical conclusion they would have to also reject their uniformitarian theory that gave them their astronomical 5 billion years old age to begin with.[154]

The rate that cosmic dust particles enter our atmosphere and settle on the surface of the earth has been determined by scientists to be basically constant. Hans Peterson has shown from his precise measures that we get approx. fourteen million tons a year. Therefore if the claim of evolunatics… er…evolutionists (typo) is true and the earth is 5 billion years old then a deposit of space dust 182 feet thick should cover the entire earth. It doesn't exist of course, anywhere, because the earth is less than 10 thousand years

---

152 Brown, *In the Beginning*, 37.

153 Brown, *In the Beginning*, 35.

154 Huse, *The Collapse of Evolution*, 21.

old. Also, NASA was quite concerned that when astronauts went to the moon they would sink into this 182 foot dust layer which should have been there. But it wasn't. It was only around an 8th of an inch.[155]

Astronauts, 182. Billions of years old, nothing.

Also the sedimentation rate of the Mississippi river, that settles into the Gulf of Mexico per year is well known- about 300 million cubic yards. But if we divide the volume of the total deposits in the Mississippi River delta by that yearly sedimentation rate, it puts the age of the delta at roughly four thousand years old.[156]

Indeed, according to the sedimentation rates of the Mississippi River and of all the other rivers from Florida, Mississippi, Alabama, Texas and Mexico that pour into the Gulf each year, if the earth was just <u>one</u> million years old (much less the *hundreds* of millions and *billions* claimed) then <u>there would be no Gulf of Mexico</u>! It would be a sand lot. The same can be said of all the other river deltas across this world. In addition, the known rate per year that river sedimentation enters the oceans and the total amount of accumulated ocean sediment also vastly contradicts these ridiculously old evolutionary ages for our planet.[157]

Are the oceans themselves hundreds of millions and even billions of years old? Not a chance...

> "Rivers carry dissolved elements such as copper, gold, lead, mercury, nickel, silicon, sodium, tin, and uranium into the oceans at very rapid rates when compared with the small quantities of these elements already in the oceans. ...far fewer than a million years' worth of metals are dissolved in the oceans. There is no known means by which large amounts of these elements can come out of solution. Therefore, the oceans must be much younger than a million years"[158]

And then there are all of those natural gas and oil deposits across the globe which we know had to be deposited rapidly by the cataclysmic

---

155 Huse, *The Collapse of Evolution*, 22-23.

156 Huse, *The Collapse of Evolution*, 23-24.

157 Brown, *In the Beginning*, 37.

158 Brown, *In the Beginning*, 38.

burial of a planets worth of vegetation, animal and marine life. (See "A World Wide Catastrophic Flood" coming up shortly) But evolutionists have insurmountable problems explaining *their* age for these deposits…

Natural gas and oil are under very high pressure in the earth. They are held down there by a fairly impenetrable cap rock. But that cap rock is permeable and calculations show that eventually the deposits will lose their pressure: in about 10,000 years. Therefore it is clearly idiotic for the evolutionists to maintain their stance that these petroleum and gas deposits have been restrained down there under the extremely high pressures we now find them, for "millions of years" without leaking out through the rock on top.

Also experiments have shown marine and vegetable material can be converted into gas and oil in a remarkably short period of time. Only 20 minutes for plant substances that were changed into petroleum under the right pressure and temperature. It took a few hours to convert wood material into coal. So the assumption by uniformitarian geologists that the transformation of petroleum, gas and coal had to have taken millions of years is misleading at best.

Those who accept the fact that all of this had to have been actually kree ate tid (created) see all of the massive coal deposits across the earth as a result of the rapid burial by the great worldwide flood (See "A World Wide Catastrophic Flood" on page 79) of all of the plant vegetation and animals. The plant remains and the consistency of the coal deposits are the obvious result of water turbulence and not some still swamp.

Evolutionists also are looking silly because exquisite jewelry and artifacts along with the skeletal remains of their human creators (☺) are found in coal deposits that supposedly were formed many millions of years before man supposedly evolved.

The rotation of the earth also stands opposed to these hundreds of millions and billions of years. It is gradually slowing because of the drag of the sun and moons gravity. If the uniformitarian geologists are right then the earth's rotation should have slowed down to zero by now.[159]

The moon's recession rate also speaks to a young age for our earth. If the earth were billions of years old then the moon would have to be at a considerably further distance. These astronomically inflated and mythological

---

159 Huse, *The Collapse of Evolution*, 24-25.

ages put forth by our evolunatic friends in order to give species time enough to have "evolved" apparently do NOT exist.[160]

Also, the amount of atmospheric helium is consistent with an age of the earth of less than 10,000 years. Helium is produced by radioactive decay within rocks and if the earth were even one million years old there would be far more helium in our atmosphere.[161] The size of the world's human population is another indicator of a relatively young earth.

And then there is the conclusive testimony of zircon crystals:

> "Lead diffuses (or leaks) from zircon crystals at known rates that increase with temperature. Because these crystals are found at different depths in the Earth, those at greater depths and temperatures should have less lead. If the Earth's crust is just a fraction of the age claimed by evolutionists, measurable differences in the lead content of zircons should exist in the top 4,000 meters. Instead, no measurable difference is found.

> "Similar conclusions are reached based on the helium content in the same zircon crystals. Because helium escapes so rapidly and so much helium is still in zircons, they (and the Earth's crust) must be less than 10,000 years old. Furthermore, the radioactive decay that produced all that helium must have happened quite rapidly, because the helium is trapped in *young* zircons."[162]

> "Based on the measured helium retention, statistical analysis gives an estimated age for the zircons of 6000 +/- 2000 years. This age agrees with literal biblical history ["Well in that case, it can't be true. Throw that science out! And didn't we tell you to burn that book?!"] and is about 250,000 times shorter than the conventional age of 1.5 billion years for the zircons. The conclusion is that helium diffusion data strongly supports the young-earth view of history."[163]

---

160   Huse, *The Collapse of Evolution*, 25-26.

161   Brown, *In the Beginning*, 37.

162   Brown, *In the Beginning*, 37.

163   Dr. Don DeYoung, *Thousands… Not Billions: Challenging an Icon of Evolution, Questioning the Age of the Earth* (Green Forest, AR: Master Books, 2006), 76.

In addition there are those gosh darn fission tracks and radiohalos. (At what point do they stop ignoring all of the evidence, or stop calling themselves scientists? What is the breaking point? Is there one?)

> "The abundance of fission tracks and radiohalos …in crystalline solids… provides evidence for a recent creation. This follows because the host rocks have not experienced serious heating since the track and halo formation. Just hundreds of degrees are sufficient to erase the crystal defects, yet they remain. It is difficult to imagine the rock formations remaining cool over vast ages of time with accompanying episodes of volcanic and tectonic activity. In the young-earth view, the radiohalos and tracks remain relatively recent and freshly made."[164]

> "Abnormally high oil, gas, and water pressures exist within relatively permeable rock. If these fluids had been trapped more than 10,000 to 100,000 years ago, leakage would have dropped these pressures far below what they are today. This oil, gas, and water must have been trapped suddenly and recently."[165]

When you take the rate that volcanoes eject material into the atmosphere each year, and the fact that only "25% of Earth's sediments are of volcanic origin" then the age of the Earth becomes only a tiny fraction of the 4.5 billion that evolutionists insist.[166]

> "The continents are eroding at a rate that would level them in much less than 25 million years. However, evolutionists believe that fossils of animals and plants at high elevations have somehow avoided this erosion for more than 300 million years. Something is wrong."[167]

Something is terribly wrong. ("Quick, we need a new story that explains away those high altitude fossils 'resistance to erosion.' Think fast.") ("See

---

164   DeYoung, *Thousands… Not Billions,* 106.

165   Brown, *In the Beginning,* 37.

166   Brown, *In the Beginning,* 37.

167   Brown, In *the Beginning,* 37.

no truth. Hear no evidence.") (Monkey see. Monkey do.) (Monkey learn. Monkey repeat.)

In sedimentary layers there are *conformities* and *unconformities*. An unconformity between layers is where you have signs of water and wind erosion, animal burrowing, and tree rooting's that show a passage of some period of time. Layers that are parallel, with a perfectly straight laser-line separating them and with no sign of un*conformities* or the passage of time, are called conformities. They "imply continuous, relatively rapid deposition."

> "Frequently, two adjacent and parallel sedimentary layers contain such different index fossils that evolutionists conclude they were deposited hundreds of millions of years apart. However, because the adjacent layers are conformable, they must have been deposited without interruption or erosion. ... Often, in sequences showing no sign of disturbance, the layer considered older by evolutionists is on top! ... Evolutionary dating rules are self-contradictory."[168]

"OK! Who's the wise guy? Who put the older sedimentary layer on top of the younger one? You've had your laugh, now put them back where they're supposed to be, and then report to the principal's office. Sheesh!"

The sodium levels in the oceans, the erosion rates of the continents and "the fact that convincing soil layers, or even soil materials, are seldom found in the geologic record" also demand a much younger earth than what Darwinists claim.[169] A massive buildup of soil layers and soil materials due to so many endless years of plant and tree activity should have occurred in between each successive sedimentary layer if indeed they are hundreds of millions of years old with millions of years passing between each successive layer. But <u>there</u> is <u>none</u>. No soil layers. No soil materials. No sign of wind or water erosion. Because absolutely no time passed between the deposition of each layer. They were all put down, one on top of the other, during the great global flood that occurred 4,500 years ago. (Again, see "A World Wide Catastrophic Flood.")

---

168 Brown, *In the Beginning,* 36-37.

169 Morris, *The Young Earth,* 89-91, 92-93, 99-100.

"Many people have the mistaken impression that geology has proved that the earth is billions of years old. As we have seen, nothing could be further from the truth! … If the Bible is really true, then the geologic evidence must support it—and indeed it does! The evidence not only supports the Bible, but a great deal of geologic evidence is quite incompatible with an old-earth scenario."[170]

"Estimated old ages for the Earth are frequently based on "clocks" that today are ticking at extremely slow rates. For example, coral growth rates were thought to have *always* been very slow, implying that some coral reefs must be hundreds of thousands of years old. More accurate measurements of these rates under favorable growth conditions now show that no known coral formation need be older than 3,400 years. A similar comment can be made for growth rates of stalactites and stalagmites in caves. … [In one recent instance] water from an underground spring was channeled to [a] spot on a river bank for only one year. In that time, limestone built up around sticks lying on the bank."[171]

…Which accumulated "millions of years" worth of deposits in just one year. Oops. Better take another look at the ages of all those caves again…

In addition to a young earth, the evidence for a relatively young universe and solar system is overwhelming and incontrovertible. The very presence of comets, which should have had their masses "boiled off" to nothing by the sun in less than 10,000 years cry out for a very young solar system. And star clusters which contain either many hundreds or even thousands of stars buzzing all around one another "like a swarm of bees" as one author said. But in a number of these clusters the movement of their stars is so fast that the notion that they have stayed together for "billions" or even "millions" of years is really dumb. Therefore the very existence of all of these clusters of stars cries out, scientifically (we are dealing with science, right? And not the opinions, beliefs, notions, biases, prejudices, brainwashing, religion and/ or utter stupidity of Darwinists and their followers…), for an age of this

---

170 Morris, *The Young Earth*, 115-118.

171 Brown, *In the Beginning*, 34.

universe that is measured in the thousands, and not billions or even millions, of years.[172]

And you're wondering why no one in the media has ever had the sense to report any of this. So am I. (Maybe we should ask Hannity, or Tucker Carlson.)

And then there are those stars which radiate their energy from a hundred thousand to even a million times faster than the sun. Millions and billions? Forget about it. They would not have had enough hydrogen to carry their fusion along at this rate. The age of those stars has to be measured in only thousands of years.

Scott Huse goes on to say that beliefs, even by a majority, doesn't make those beliefs true. Arguing that the majority believe something is irrelevant to scientific truth. Because you don't arrive at that truth by counting votes. Just how many times have majorities been totally wrong?[173]

(Just look at the 2008 presidential and 2018 congressional elections...) (I mean, for God's sake, the idiots elected a certified moron, Ocrazyo Cortez.)

The evidence from meteorites is also on the side of a relatively young earth and solar system...

> "Experts have expressed surprise that meteorites are almost always found in young sediments, very near Earth's surface. Even meteoric particles in ocean sediments are concentrated in the topmost layers. If Earth's sediments, which average about a mile in thickness on the continents, were deposited over hundreds of millions of years, as evolutionists believe, we would expect to find many deeply buried iron meteorites. Because this is not the case, the sediments were probably deposited rapidly, followed by "geologically recent" meteorite impacts.
>
> ..."Similar observations can be made concerning ancient rock slides. Rock slides are frequently found on Earth's surface, but are generally absent from supposedly old rock."[174]

---

172 Huse, *The Collapse of Evolution*, 29-30.

173 Huse, *The Collapse of Evolution*, 30-31

174 Brown, *In the Beginning*, 38.

Also, the present temperature of the earth, the recession rate of the moon, the "amount of heat …flowing out of the Moon from just below its surface, [although] the Moon's interior is relatively cold," and the lack of hundreds of millions or billions of years' worth of meteoric material on the moon all point to a very young earth-moon system. There is also the absence of any signs of "flattening out" (called "creep") of the large, steep-walled craters on either the Moon, Venus or Mercury (which should have happened if they were formed over 4 billion years ago as evolutionists say). All of these are further strong scientific evidence of youth and not great age.[175]

> "As comets pass near the Sun, some of their mass vaporizes, producing a long tail and other debris. Comets also fragment frequently or crash into the Sun or planets. Typical comets should disintegrate after several hundred orbits. For many comets this is less than 10,000 years. There is no evidence for a distant shell of cometary material surrounding the solar system, and there is no known way to add comets to the solar system at rates that even remotely balance their destruction. Actually, the gravity of planets tends to expel comets from the solar system rather than capture them. So, comets and the solar system appear to be less than 10,000 years old."[176]

> "Jupiter, Saturn, and Neptune each radiate away more than twice the heat energy they receive from the Sun. Uranus and Venus also radiate too much heat. Calculations show that it is very unlikely that this energy comes from nuclear fusion, radioactive decay, gravitational contraction, or phase changes within those planets. This suggests that these planets have not existed long enough to cool off."

> "The Sun's radiation applies an outward force on particles orbiting the Sun. Particles less than about one 100,000th of a centimeter in diameter should have been "blown out" of the solar system if it were billions of years old. Yet these particles are still orbiting the Sun."[177]

---

175 Brown, *In the Beginning,* 39.

176 Brown, *In the Beginning,* 39-40.

177 Brown, *In the Beginning,* 40

And concerning those particles which are <u>larger</u> than one 100,000th of a centimeter, there is "a large disk shaped cloud that orbits the Sun between... Venus and the asteroid belt. [It consists] of dust particles larger than about one 100,000th of a centimeter in diameter." However, the well-established *Poynting-Robertson effect* states that "forces acting on these particles should spiral most of them into the sun in less than 10,000 years. ...Known forces and sources of replenishment cannot maintain this cloud, so the solar system is probably less than 10,000 years old."[178] Probably.

> "Huge quantities of microscopic dust particles also have been discovered around some stars. Yet, according to the theory of stellar evolution, those stars are many millions of years old, so that dust should have been removed by stellar wind and the Poynting-Robertson effect. Until some process is discovered that continually resupplies vast amounts of dust, one should consider whether the "millions of years" are imaginary."[179]

Uh, we've already considered it. Trust me, they're imaginary.

> "In galaxies similar to our Milky Way Galaxy, a star will explode violently every 26 years or so. These explosions, called *supernovas*, produce gas and dust that expand outward thousands of miles per second. With radio telescopes, these remnants in our galaxy should be visible for a million years. However, only about 7,000 years' worth of supernova debris are seen. So, the Milky Way looks young."[180]

The Milky Way looks <u>real</u> young. And the lies concerning these ridiculous and astronomically inflated ages for the earth and the solar system that they have concocted in order to support the absurdity that is Satan's genesis, that they are indoctrinating our children with, are looking rather sick.

In addition there are: galaxies that are connected yet "have vastly different redshifts;" the existence of so many spiral galaxies whose internal dynamics cannot allow them to maintain their highly esthetic shape for

---

178 Brown, *In the Beginning*, 40.

179 Brown, *In the Beginning*, 40.

180 Brown, *In the Beginning*, 40

even "a small fraction of the universe's assumed evolutionary age;" and the existence of "many stars in spiral galaxies and gas clouds that surrounds some galaxies [that] have such high relative velocities …they should have broken their "gravitational bonds" long ago if they were billions of years old." These are also strong astronomical evidence for a very young universe.[181]

> "These observations have led some to conclude, not that the universe is young [shocker!], but that unseen, undetected mass is holding these stars and galaxies together. For this to work, the hidden mass, sometimes called *dark matter*, must be 10 - 100 times greater than all visible mass, and the hidden mass must be in the right places. However, many experiments have shown that the needed "missing mass" does not exist. Some researchers are still searching, because the alternative is a young universe."[182] [O. J. is still searching too. He couldn't find his wife's killer out on the golf course, so now he's looking in prison.] [Are there mirrors in prison?]

If these folks were actual scientists, they might come up with a different theory to explain these observations, one that doesn't include a tooth fairy called "missing mass." But as high priests of a false religion, they want to protect their sacred doctrine at all costs, even at the expense at their own scientific credibility. In every other scientific discipline, all of those which have no relevance to evolution or the Bible, when observations can't be fit into existing theories it is the theory that goes. Instead evolutions high priests will conjure up any explanation no matter how laughable ("Here, missing mass. Here, boy. Tweet, tweet, tweet. Come out, come out, wherever you are."), rather than even entertaining the idea that their astronomically inflated timeframes don't exist.

- Mitochondrial Eve In 1987 three scientists did a study of mutations within the DNA of the mitochondria, which are energy-producing units in the cell and that come only from the mothers DNA. They studied 147 people from all over the world and determined that all of them had the same ancestral mother. There mistake however was that they used the *assumption*

---

181 Brown, *In the Beginning*, 41.

182 Brown, *In the Beginning*, 41.

of evolution in determining the mutation rate of the mtDNA (mitochondrial DNA) and thus how long ago the first woman lived…

> "**When** did mitochondrial Eve live? To answer this, one must know how frequently mutations occur in mtDNA. Initial estimates were based on the following *faulty* reasoning: "Humans and chimpanzees had a common ancestor about 5 million years ago. Because the mtDNA in humans and chimpanzees differ in 1000 places, one mutation occurs every 10,000 years." …These estimated rates, based on evolution, led to the mistaken belief that mitochondrial Eve lived 100,000-200,000 years ago. [But even] this surprised evolutionists who believe that our common ancestor was an apelike creature that lived 31/2 million years ago."

[So even when using their faulty circular reasoning to determine the mutation rate of mtDNA, they still came up with an age for the first woman that was 30 times less than what they needed for their theory. But it gets even worse…]

> "A greater surprise, even disbelief, occurred in 1997, when it was announced that mutations in mtDNA occur 20 times faster than had been estimated. Without assuming that humans and chimpanzees had a common ancestor 5 million years ago… mutation rates can now be determined directly by comparing the mtDNA of many mother-child pairs. Using the new, more accurate rate, ***mitochondrial Eve lived only about 6,500 years ago.***"[183]

<u>Bingo</u>! So science confirms the time line of roughly 6,000 years ago that the Word of God gives through its genealogical record for the first woman, Eve, and one would assume therefore for the first man, Adam. Isn't science a wonderful thing?

<u>- Radioactive dating techniques</u> But what about all of those scientific dating methods like carbon-14 and the radioisotopes of rocks; aren't they *conclusive evidence* for the earth being <u>hundreds</u> of millions and billions of years old? No, not hardly.

---

183   Brown, *In the Beginning,* 319-320.

First, let's look at carbon-14:

> *"For some years there has been a growing realization that carbon-14 atoms are found where they are not expected.* With a half-life of 5730 years, C-14 should no longer exist within "ancient" fossils, carbonate rocks, or coal. Yet small quantities of C-14 are indeed found in such samples on a worldwide scale. ...Measurable levels of C-14 are found in every case for both coal and diamond samples. This evidence supports a limited age for the earth. There is a widely held misconception that carbon-14 dating is in direct conflict with creation and a young-earth view. Instead, however the carbon-14 findings strongly support a recent supernatural creation."[184]

Whoops! There should be no carbon-14 whatsoever in all of the rocks and sediments and deposits that are supposedly more than a million years old, much less hundreds of millions of years old! But there they all are!

> "To understand the significance of this carbon-14 finding, consider a comparison. Suppose an archaeologist investigates an Egyptian mummy. The outer covering is carefully removed to reveal the ancient, undisturbed interior. As the last wrapping is removed an amazing discovery is made. Inside the mummy is a wind-up clock which is still ticking! Perhaps the mummy is not as old as the archaeologist initially thought. The discovery of carbon-14 in "ancient" samples is just as startling to the conventional radio isotope dating community.

[Must have been real startling, because they've already ignored it. For health reasons of course.]

> ..."The presence of C-14 in "very old" fossils, rocks, coal, and diamond samples is clearly in major conflict with the long age time scale... The conclusion is that the pervasive presence of C-14 is strong evidence for a young earth."[185]

---

184   DeYoung, *Thousands... Not Billions,* 175-176.

185   DeYoung, *Thousands... Not Billions,* 49-50.

Radioisotope dating, that which has been used for decades to "prove" that the earth is billions of years old, has recently been proven to be completely unreliable due to a number of false assumptions.[186] Here is just one glaring discrepancy, the vast difference between the *known* ages of certain rocks and their *radioisotope age*:

> "We can gather samples, for example, from recent volcanic eruptions and date them. If the dating process is accurate, then the date derived should be almost equivalent to zero, or too young to be measured. In the scientific literature, research results have often been reported where rocks of known ages have been dated. In almost every case, the age of these recent lavas has come back from the lab in terms of excessively high ages, not essentially zero, as one would predict.

> "Let me give you a few examples. It is known that Sunset Crater in northern Arizona was formed by a series of recent volcanic eruptions. ...Tree-ring dating accurately dates the eruption to about A.D. 1065, and it is instructive to compare this historical date with the radioisotope date obtained.

> "Two of the lava flows were dated by the potassium-argon method. Much to everyone's surprise, the lava flows gave the "ages" of 210,000 and 230,000 years!

> ..."Consider another example. A coal mine in Queensland, Australia, required a vertical ventilation shaft to provide air to the miners. On the way down the drill encountered a basalt layer, and underneath the basalt they found pieces of unfossilized wood. Multiple carbon dating studies of the wood fragments yielded an "age" of 30,000 to 45,000 years, while the basalt, using the potassium-argon method, dated 39-58 million years!

> "[Again, rocks were dated] from a volcano in ...New Zealand, Mt. Ngauruhoe. It has often erupted in recent decades, and it was these recently formed rocks that were gathered and dated by multiple methods. While K-Ar model ages dated from 270,000 years to 3.5 million, the Rb-Sr

---

186   (For a complete discussion you can see Chapter 5, "Radioisotope Dating," from Dr. John Morris' book, *The Young Earth: The Real History of the Earth – Past, Present, and Future.*)

isochron dated over 133 million years, the Sm-Nd method nearly 200 million years and the lead-lead ratio method indicated a date of **3.9 billion years**! [Emphasis mine.] All this from rocks less than 60 years old. Do radioisotope-dating results warrant our trust?"[187] (Absolutely! If we are complete morons.)

Wow! 3.9 billion years old! From rocks less than a hundred years old. (And they called Sarah Palin stupid!) And there are countless other examples of volcanoes that have erupted recently and yet yield astronomically inflated ages.

The evolutionists say that diamonds were formed billions of years ago in earth's "earliest time" which means that it is absolutely impossible for these crystal forms of carbon to contain any C14. OOPS! Recent tests of diamonds from a number of various places have shown that they contain measurable amounts of C14 which means that these diamonds formed only thousands of years ago, and not billions.[188]

But please don't tell anyone! We don't want to expose that little man behind the curtain. And then there's this little known fact, that the ages of igneous rocks on the Grand Canyons rim, by means of radioisotope dating are OLDER than igneous rock sediments on the Canyons floor.[189]

Oops, now how did *that* fly get in the ointment? (But Darwinists have an answer. They got away with convergent evolution so now, emboldened by that success, they've proposed *reverse* sedimentation, where the most recent sedimentary layers actually "fall" *below* the older ones. Shazaam!)

In addition, the very assumptions that radioisotope dating are based upon are themselves questionable…

"The public has been greatly misled concerning the consistency and trustworthiness of radiometric dating techniques (such as the potassium-argon method, the rubidium-strontium method, and the uranium-thorium-lead method)…

"A major assumption underlying all radioactive dating techniques is that decay rates, which have been

---

187   Morris, *The Young Earth*, 51-52.

188   Morris, *The Young Earth*, 67.

189   Morris, *The Young Earth*, 78.

essentially constant over the past 100 years, have also been constant over the past 4,600,000,000 years. This is a huge and critical assumption that few have questioned. [And since Darwinists depend on that assumption being true, why would they question it?] Several lines of evidence show that radioactive decay rates were once much faster than they are today."[190]

And then there is DNA "the immortal"!

"When an animal or plant dies, its DNA begins decomposing. Before 1990, almost no one believed that DNA could last 10,000 years. This limit was based on measuring DNA disintegration rates in *well-preserved* specimens of known age such as Egyptian mummies. DNA has now been reported in supposedly 17-million-year-old magnolia leaves and 11-to-425-million-year-old salt crystals. ...DNA fragments are also said to be in alleged 80-million-year-old dinosaur bones buried in a coal bed and in the scales of a 200-million-year-old fossilized fish. DNA is frequently reported in insects and plants encased in amber, both assumed to be 25-120 million years old.

[These discoveries alone should totally discredit these asinine ages, but religious fanatics will fight against *heresy* (in this case the *truth*) even to the death! And in this case the death of the scientific method.]

"These discoveries have forced evolutionists to re-examine the 10,000 year limit [of DNA]. [Shocker! Curiously, they never stopped to re-examine their fraudulently old time scales.] They now claim that DNA can be preserved longer if conditions are dryer, colder, and freer of oxygen, bacteria, and background radiation. However, measured disintegration rates of DNA, under these more ideal conditions, do not support this claim."[191] ("Hear no truth. See no evidence.")

---

190 Brown, *In the Beginning*, 34.

191 Brown, *In the Beginning*, 35-36.

Live bacteria spores have been found in bees preserved in amber that is dated to 25 to 40 million years old. They have also been found in rocks dated 250 to 650 million years old. And "proteins, soft tissues, and blood compounds [have been found] preserved in dinosaur bones." None of which could last for even a tiny fraction of the supposed ages of the material in which they were found. Human artifacts such as "a thimble, an iron pot, an iron instrument, an 8-karat gold chain, three throwing spears, and a metallic vessel inlaid with silver" have been discovered in coal deposits that are, according to the "established" dates, *300 to 350 million years old!* In addition "nails, a screw, a strange coin, a tiny ceramic doll" and other artifacts were found inside of deeply buried rock, again *dated* to hundreds of millions of years old. Yet according to the evolutionists' time scales, man is less than <u>one</u> million years old! (And what do they say when presented with such obvious and incontrovertible contradictions? Nothing. They just ignore them.) Obviously, these dates are absurd. This amber, these rocks, those dinosaurs, that coal and this earth are all very young, at the maximum about 10,000 years old. The Bible again is found to be scientifically accurate, and the dates claimed by the uniformitarian evolutionists are found to be *astronomically* exaggerated to fit their preconceived and intransigent worldview.[192]

"For in six days the Lord made the heavens and the earth, the sea, and all that in them is." (Exodus 20:11)

As we can see the evidence is staggering. It is overwhelming. (And we have only presented a tiny fraction of it here.) The earth is very young. Why this isn't common knowledge is due to the fact that we live in an insane asylum, and the inmates have truly taken over. They own our schools, our entertainment industry, our late-night clowns and of course our media (unfortunately including Fox News: last night on the Gutfeld show Greg again used his ignorant, uneducated belief in evolution as a tidbit in his monologue), and their unscientific lies cover all of it like a shroud, and people believe what they have been taught.

Again, as we stated in chapter 6, "The Word of God," (in the larger work) the Bible says that the universe was <u>created</u> about six thousand years ago. Faux science says that the universe sprung up billions of years ago from a big explosion and then formed itself, all that we survey today, accidentally by random chance. The Biblical account is scientifically accurate. The explosion and random chance account is scientifically absurd. And the unscientific

---

192   Brown, *In the Beginning,* 36.

notion that the universe is billions of years old did not arise from unbiased observations or a dispassionate compilation of the facts and evidence in order to ascertain their meaning. Instead it came from a biased search that had at the outset what the "find" was going to be, astronomically old ages that could be used to "validate" the ridiculous theory of evolution. And to keep selling their find today it is necessary to distort the evidence, ignore the observations, cover up the facts, and to deliberately overlook the truth. What any of this has to do with actual science, you'd have to ask them.

Starlight and Time

"Bless the Lord, O my soul! O Lord my God, you are very great; you art clothed with honor and majesty. Who covers yourself with light as with a garment. Who stretches out the heavens like a curtain. Who …makes the clouds his chariot, who walks upon the wings of the wind." (Psalms 104:1-3)

Both scientifically and biblically we know that the universe is less than 10,000 years old. But a major question still remains, *how did the light from all of those distant galaxies, some as far as twelve billion light-years away, get here so quickly?* And it's a very good question. Light travels at 186,000 miles per second, which is 5,865,696,000,000 miles every year. A distance of 12 billion light years is 70,388,352,000,000,000,000,000 miles away. (Or roughly the size of our National Debt shortly after the Socialist, lying, baby-mutilating Democrats get elected.) That is also the distance light must travel, going extreeeemely quickly, in order to arrive here 12 billion years later from that distant galaxy 12 billion light-years away. However, if the entire universe is only around 6 <u>thousand</u> years old, which it of course is, then how did that light get here already? It should still be 70,388,316,805,824,000,000,000 miles away, or just slightly farther than Tina Fey can see out of her bathroom window.

The answer lies in exactly <u>how</u> God made the universe. It turns out, as we mentioned before in the section on "The Big Bang," that he "stretched out the heavens like a curtain." (Isaiah 40:22) He created matter and then formed the stars and the galaxies and then stretched them out to their present positions. And the beams of light from those stars and galaxies were stretched out as well. So the light never had to "get here" to earth. It was here from the beginning. This "stretching out" or expansion also caused the redshifts of distant stars and galaxies. The following was quoted earlier in "The Big Bang" section but it bears repeating here:

> "A... misconception is that the red shifts of the galaxies are Doppler shifts, i.e. caused by the velocity of the galaxies away from us at the time the light starts its journey toward us. But one undergraduate textbook... and many graduate textbooks... make it clear that the red shifts are an *expansion* effect. As space is stretched out, the lengths of all electromagnetic waves passing through the space are similarly stretched out. Consequently, the *speed* of recession doesn't matter, only the *amount* of expansion that takes place as the light travels to us, whether the expansion is fast or slow."[193]

"He has made the earth by his power, he has established the world by his wisdom, and has stretched out the heavens by his discretion." (Jeremiah 10:12)

This "stretching out" or expansion is the cause of both the redshifts of distant stars and galaxies, and the reason, along with gravitational time dilation,[194] why we can still see the light from those distant galaxies (some as far as 12 billion light years away) even though the earth and our universe are very young, less than 10,000 years old.

But what about the Dinosaurs?

"Behold now the behemoth, which I made along with you; he eats grass like an ox. See now, his strength is in his hips, and his power is in his stomach muscles. He moves his tail like a cedar; the sinews of his thighs are tightly wrapped together. His bones are as strong beams of brass, his ribs are like bars of iron." (Job 40:15-18)

Evolutionists tell us that Dinosaurs roamed the earth hundreds of millions of years ago. They say they *know* this because their fossils are found in the "geologic column" during this time period, and that they have been extinct for about the same amount of time. But all that of course is a lie. We know now that the sedimentary layers that make up the Darwinists imaginary geologic column are not hundreds of millions or billions of years old, but only thousands of years. We also know that their geologic column is a fabrication;

---

193   Humphreys, *Starlight and Time*, 98.

194   (For an explanation of "gravitational time dilation" see pgs. 11-13 of *Starlight and Time: Solving the Puzzle of Distant Starlight in a Young Universe*, by D. Russell Humphreys, Ph.D. We would explain it here, but it gives us a headache)

it does not exist in reality, nor in the earth's sediments; nor in real science; it only exists in their own minds and imaginations.

The scientific evidence actually points to these Dinosaur creatures existing alongside of "modern" man just in the recent past. We mentioned in a previous segment examples of fossilized human footprints being found next to Dinosaur prints, but it bears revisiting…

> "Scientists in the former Soviet Union have reported a layer of rock containing more than 2,000 dinosaur footprints alongside tracks resembling human footprints. Obviously, both types of footprints were made in mud or sand that later hardened into rock. If some are human footprints, then man and dinosaurs lived at the same time. Similar discoveries have been made in Arizona. Were it not for the theory of evolution, few would doubt that these were human footprints. [But hey, what are you going to believe, evolution or your lying eyes?]

> "Soft tissue has now been recovered from several dinosaurs: three tyrannosaurs (*T. Rex*) and one hadrosaur. It is ridiculous to believe that soft tissue can be preserved for more than 60 million years, but it could be preserved for 5000 years."[195]

It is also ridiculous to believe that a genital-mutilating, genocidal maniac and filthy child molester is the prophet of God, but that is the nature of false religions: people believe whatever they are taught. And if Dinosaurs are supposed to be 60 million years old, well then that's that, soft tissue preservation be damned.

Also, both human and dinosaur prints have been found together, after being exposed by construction or flood, in Arizona, Illinois, Kentucky, New Mexico, Mexico, Missouri and throughout the US. They are not "frauds." Respected paleontologists have studied these and determined their authenticity. Plus dinosaur drawings have been found on canyon rocks and in caves in Rhodesia and Arizona. Hello! Man and dinosaurs lived at the same time![196]

But if Dinosaurs lived alongside of man on this earth in the recent past shouldn't we expect to find stories and legends and historical writings about

---

195 Brown, *In the Beginning,* 350.

196 Huse, *The Collapse of Evolution,* 16-17.

them? Yes, we should by gosh by golly. And we do! Thousands of them! They're called <u>Dragons</u>! (Bingo!) You see, these "ancient" Dinosaurs and their *mythological* counterpart, Dragons, are one and the same.

> "From countries all over the world, we read and hear stories of ancient dragons. Some flew, while others lived in caves or murky swamps. Some of the dragons protected emperors and cities, but most of the dragons ravaged the countryside killing animals and people.
>
> "In far-eastern countries such as China, dragons often are found in the ancient writings. Some of them were even used to pull the chariots of Chinese rulers. Also, many of the ancient Chinese people used "dragon bones" for special medicines and potions. While visiting the continent in the 1200s, the Italian explorer Marco Polo said that he saw long reptiles called Lindworms that easily ran as fast as a horse!
>
> "In the British Isles, **hundreds** of dragon stories have come down to the present day. [Emphasis mine] One legend told of an animal with a crested head, teeth like a saw, and a long tail. Also, in 1449 in England, it is reported that two huge reptiles were seen fighting on the banks of the river Stour.
>
> ..."Countries all over the world have stories of dragon slayers."[197]

Really? Could it be?! Could "Dragon" and "Dinosaur" actually be two words for the same creatures?! Of course! What else could they be? It's only because of the widespread acceptance of the idiotic lie of evolution and these hundreds of millions and billions of years/imaginary time-spans, that Darwinists have been able to pull off the scam of calling Dinosaurs ancient, and Dragons a "myth." It is truly a wonder that they have been able to keep so many people in the dark for so long. The look of wonder and enlightenment (like a bright light was suddenly switched on inside of their head) on a person's face when they first learn that Dinosaurs and Dragons are one and the same is absolutely priceless. It's like, "Yea! Wow! Why didn't I think of that before? It's so obvious now." ("I could've had a V8!") We didn't think of it because we have all been lied to constantly in our schools, in our media and in our

---

197 Butt and Lyons, *Dinosaurs Unleashed*, 64-65, 75.

entertainment industry, and we believe what we are taught, and we become lost in the powerful blinding grip of a paradigm that, as completely false and ridiculous as it is in reality, can still blind us from seeing the obvious truth. (Just ask the Swiss why they sold the battery-operated watch patent for a song and a dance, and gave away what became an astronomically lucrative market to the Japanese. Pair of dimes.)

"The ancient legends of dragons – huge reptilian beasts with long necks and plated spines – depict beasts very similar to our modern reconstruction of dinosaurs. Mariners have told of dragons in the oceans even up until recent times. Almost every culture around the globe has legends of dragons, and these legends show striking similarities.

"Some dragon stories recorded in human history involve people who actually did exist. Alexander the Great encountered dinosaur-like beasts, and so did King Beowulf of the British Isles. The ancient historian Berosus wrote of such beasts. Clearly, something that fits our modern ideas of dinosaurs existed in human history… **the legends of dragons are actually the faded memories of human encounters with dinosaurs.** [Emphasis mine]

[There are accounts in the Middle Ages of valiant knights battling dragon-like beasts. And although rare, dragons are even listed in the scientific logs of animals then, and as creatures that were ALIVE! Also throughout the world- in China, Africa, North and South America, Europe and Australia- there are cave and rock drawings depicting images of dinosaurs.][198]

"Consider the many dragon legends. Most ancient cultures have stories or artwork of dragons that strongly resemble dinosaurs. *The World Book Encyclopedia* states that:

The dragons of legend are strangely like actual creatures that have lived in the past. They are much like the great reptiles [dinosaurs] *which inhabited the earth long before man is supposed to have appeared on earth. Dragons were generally*

---

198   Morris, *Dinosaurs, the Lost World, & You,* 25, 33.

> *evil and destructive. Every country had them in its mythology.*
>
> The simplest and most obvious explanation for so many common descriptions of dragons around the world is that man once knew the dinosaurs."[199]

And because these Drago-saurs did not go extinct in some imaginary age 200 million years ago, but only disappeared in the last few thousand years, there is a very good possibility, and some evidence as well, that a few may still be alive today.

> "For the past three centuries, reports have come from the Congo in western Africa that dinosaurs exist in remote swamps. Eyewitness stories are often from educated people who can quickly describe dinosaurs. Two expeditions to the Congo, led by biologist Dr. Roy Mackal of the University of Chicago, never saw dinosaurs, but interviewed many of these witnesses and concluded that their reports were about dinosaurs and were apparently true."[200]

And then there is the most important witness of all, and the last one Mr. Evolutionist would ever think of entertaining, and that is the Word of God. We know that the Bible's witness is true. (And if you doubt the truthfulness of its witness we would encourage you to check out chapter 6, "The Word of God" in the larger book, "All the Secrets.") So what does it say about Dinosaurs and Dragons? Well, in the Book of Job we find some fascinating verses…

"Behold now the behemoth, which I made along with you; he eats grass like an ox. See now, his strength is in his hips, and his power is in his stomach muscles. **He moves his tail like a cedar**; the sinews of his thighs are tightly wrapped together. His bones are as strong beams of brass, his ribs are like bars of iron." (Job 40:15-18)

Job goes on to describe how he lived in swamps and was unconcerned with a rampaging river. "Indeed the river may rage, yet he is not disturbed; he is confident, though the Jordan should gush against his mouth." (Job 40:23) Sounds like a Brontosaurus[201] to me. This behemoth could <u>not</u> have been

---

199 Brown, *In the Beginning*, 350.

200 Brown, *In the Beginning*, 350-351.

201 (See "The Flintstones." …It's now called a *Brachiosaurus*)

an elephant or a hippopotamus because they both "have tails like ropes... [Yet] any animal with a tale as huge and strong as a cedar tree is probably a dinosaur."[202] Probably.

But what of *fire-breathing* dragons? Legends abound from across the globe of a Tyrannosaurus Rex type of creature who could shoot fire from his mouth and nostrils. (Of course, after putting together the bombardier beetle this was just an afterthought for the Creator. He had some leftover explosive material lying around and didn't want it to go to waste...) Could these "legends" be actual true accounts of real dinosaurs that plagued man for centuries? Let's look at the next chapter of Job...

"Can you draw out Leviathan with a hook, or snare his tongue with a line which you let down? Can you fill his skin with harpoons? Or his head with fishing spears? None is so fierce that he would dare stir him up. Who can open the doors of his face, with his terrible teeth all around? His rows of scales are his pride, shut up tightly as with a seal. One is so near another that no air can come between them. They are joined one to another; they stick together and cannot be parted. **His sneezings flash forth light... Out of his mouth go burning lights, and sparks of fire shoot out. Out of his nostrils goes smoke, as out of a boiling pot or caldron. His breath kindles coals, and a flame goes out of his mouth.** When he raises himself up, the mighty are afraid; because of his crashing about they are beside themselves. Though the sword reaches him, it cannot avail; nor does the spear, dart, or javelin. He regards iron as straw, and brass as rotten wood. The arrow cannot make him flee; sling stones are turned into stubble by him. Darts are regarded as straw; he laughs at the threat of spears. His undersides are like sharp stones." (Job 41:1, 7, 10, 14-21, 25-30)

If old Job wasn't describing dinosaurs that coexisted with him and his contemporaries then he must have gotten a bootlegged copy of *Jurassic Park*. There are legends of fire-breathing dragons, thought by the evolutionists to be myths, but there we see Job actually describing one! And we know today that a fire-breathing dragon is possible because the chemistry required for this ability is actually less complicated than what is already in the possession today of animals like the electric eel, the firefly, etc. Due to the heavy plant diet of most dinosaurs, and the ignitable methane gas that would be produced, all that is then required is just a spark. That spark could be provided by a number

---

202   Brown, *In the Beginning,* 350.

of known chemicals which, in the company of oxygen and methane erupt into flame. The end result would be an awesome flame thrower.[203]

The truth about Dinosaurs is they are Dragons. And they lived and breathed and walked the earth with man since the dawn of creation six thousand (and not hundreds of millions of) years ago. They didn't *evolve* from any other organism, nor did any other organisms evolve from them. The Bible is absolutely accurate, and the theory of evolution is a joke. (Somebody might want to tell The New York Times.) (And Fox News.) (Gutfeld? Tyrus? Are you paying attention??)

A World Wide Catastrophic Flood

"In the six hundredth year of Noah's life, in the second month, the seventeenth day of the month, on that day were all the fountains of the great deep broken up, and the windows of heaven were opened. And the rain was upon the earth forty days and forty nights." (Genesis 7:11-12)

The overwhelming evidence from the Earth Sciences indicate that a worldwide flood did occur in the recent past, just as the Bible records. And this evidence clearly shows, among many other things, that…

> …"coal, oil, and methane did not form over hundreds of millions of years; they formed in months. Fossils and layered strata did not form over a billion years; they formed in months. The Grand Canyon did not form in millions of years; it formed in weeks. Major mountain ranges did not form over hundreds of millions of years; each formed in hours."[204]

Before we get to all of the facts, though, there is the historical evidence from the many so-called "myths" that all tell of a catastrophic deluge and can be found in nearly every culture:

> "A gigantic flood may be the most common of all legends—ever. Almost every ancient culture had legends telling of a traumatic flood in which only a few humans survived in a large boat. This cannot be said for other types of catastrophes, such as earthquakes, fires, volcanic

---

203   Morris, *Dinosaurs, the Lost World, & You*, 36-37.

204   Brown, *In the Beginning*, 103.

eruptions, disease, famines, or drought. More than 230 flood legends contain many common elements, suggesting they have a common historical source that left a vivid impression on survivors of that catastrophe."[205]

There are legends of creation shared by numerous peoples all across this earth that are remarkably similar. They tell of a time when there was only one language, an abundance of food and long lifespans, but this golden era ended in a huge flood because they disobeyed their Creator (which BTW, was NOT random accidental chance) (jus sayin). Other common legends, not as numerous, talk of how God scattered the tribes and they migrated and reestablished various civilizations all around the globe. Even more incredible is how anthropologists have found shocking resemblances between almost all cultures folklore. Many hundreds of peoples scattered all across the earth talk of God sending a huge flood to destroy man and his wickedness, but one upright family was saved by building a large boat which settled, after the waters subsided, on a great mountain. Wow! One might actually be led to think, if one wasn't evolutionary indoctrinated, that because of this common memory, all of us must be descendants of a common ancestor, Noah. [206]

God's word is true. Every part of it. (For absolute proof of that, see chapter 6, "The Word of God" in "All the Secrets.) And the common legends of a world-wide flood from every corner of humanity bear witness to it.

The Bible talks of "all the fountains of the great deep" being broken up in one day. (See Genesis 7:11 above.) Those words are the key to understanding exactly what happened to cause this great, worldwide flood. When God created the earth there were vast reservoirs of water under the earth's crust. This is borne out by ample physical evidence all across the globe that came from the aftermath of Noah's Flood, but let's also see what the Bible says:

"And God made the firmament, and divided the waters which were under the firmament from the waters which were above the firmament; and it was so. And God called the firmament Heaven. And the evening and the morning were the second day." (Genesis 1:7-8)

An accurate interpretation of the word *firmament* (firm) in this verse is not *sky*, but earth, or more accurately, the earth's *crust*. Also in this verse when

---

205　Brown, *In the Beginning,* 41.

206　Morris, *The Young Earth,* 73.

God calls the firmament *Heaven* he was not referring to the sky or to outer space or to the place where He dwells, he was referring to the paradise here on earth that he had created originally before the fall of man and the subsequent curse.[207] So there is both ample evidence from the earth and from the Bible that there were these vast subterranean caverns of water, these waters of the "great deep," held under the earth's crust at enormous pressure, that, when released, produced a catastrophic worldwide flood.

> "New evidence shows that the earth has experienced a devastating, worldwide flood, whose waters violently burst forth from under the earth's crust. Standard "textbook" explanations for many of the earth's major features are scientifically flawed. We can now explain, using well-understood phenomenon, how this cataclysmic event rapidly formed so many features."[208]

These features—which include the Grand Canyon, the mid-oceanic ridges, continental drift, continental shelves and slopes, the ocean trenches and the ring of fire, earthquakes, magnetic variations on the ocean floor, submarine canyons, coal and oil deposits, methane hydrates, the Ice Age, frozen mammoths, major mountain ranges, overthrusts, volcanoes and lava, geothermal heat, strata and layered fossils, limestone, metamorphic rock, plateaus, the Moho, salt domes, jigsaw fit of the continents, comets, asteroids, and meteorites—are all…

> …"a consequence of a sudden, unrepeatable event—a global flood, whose waters erupted from interconnected, worldwide subterranean chambers with an energy release exceeding the explosion of 300 trillion hydrogen bombs."[209]

Dr. Walt Brown explains in his book, *In the Beginning: Compelling Evidence for Creation and the Flood*, how this water was released along a rapidly-spreading, north-south crack in the earth's crust above the line of what is now the Mid-Atlantic Ridge, and then spread globally along the Mid-Oceanic Ridge, which "wraps around the Earth and is the world's longest

---

207   (For an in-depth discussion of this see pgs. 365-370, *In the Beginning: Compelling Evidence for Creation and the Flood,* by Walt Brown, Ph.D.)

208   Brown, *In the Beginning,* 105.

209   Brown, *In the Beginning,* 105.

mountain range—46,000 miles."[210] This water was under so much pressure, 62,000 psi, that it quickly eroded the sides of this ever-widening split in the earth's crust, and shot tremendous volumes of water and earthen debris into the atmosphere and out into outer space.

> "When the flood began, the pressure in the jetting SCW [supercritical water][211] dropped in seconds from at least 62,000 psi (4,270 bars) to almost zero. The energy released was huge. Because the 46,000-mile-long fountains continued this release for several weeks, one should not think of it as a single explosion. Instead, the jetting water was a powerful, earth-size engine that launched considerable mass from earth."[212]

> "As the crack raced around the earth, the 10-mile-thick crust opened like a rip in a tightly stretched cloth. Pressure in the subterranean chamber directly beneath the rupture suddenly dropped to nearly atmospheric pressure. This caused supercritical water to explode with great violence out of the 10-mile-deep slit that wrapped around the earth like the seam of a baseball.

> "All along this globe-circling rupture, whose path approximates today's Mid-Oceanic Ridge, a fountain of water jetted supersonically into *and far above* the atmosphere. Much of the water fragmented into an "ocean" of droplets that fell as rain great distances away. This produced torrential rains such as the earth has never experienced--before or after."[213]

This breaking up of "all the fountains of the great deep" was also the origin of comets, asteroids and meteoroids, as much water along with mass quantities of rock debris from the rapidly eroding sides of the great "globe-circling rupture," had enough velocity to escape earth's orbit.

---

210   Brown, *In the Beginning,* 105.

211   Brown, *In The Beginning,* 118. (As the pressure on water increases, so does the temperature at which it will boil. When water reaches pressures above 3,200 psi (called the critical point) it becomes supercritical and cannot boil.)

212   Brown, *In the Beginning,* 119.

213   Brown, *In the Beginning,* 120-121.

> "The most powerful jetting water and rock debris escaped the earth's gravity and became the solar system's comets, asteroids, and meteoroids."[214]

Again, in the beginning God created a perfect world, without all of this solar debris orbiting the sun and threatening to hit earth with a large enough chunk to wipe out all of life. That's not what God had in mind when he finished Creation and "saw every thing that he had made, and, behold, _it was very good_. And the evening and the morning were the sixth day." (Genesis 1:31) The very bad came a little later…

> "When the fountains of the great deep erupted, rocks were crushed, eroded, and sometimes reduced to clay. Mixed with that debris was carbonate-rich, salty, subterranean water... Organic compounds, including methane and ethane, are found in comets, because that water contained pulverized vegetation from preflood forests (as well as bacteria and other traces of life) from within hundreds of miles of the globe-encircling rupture."[215]

> These "strange bodies, sometimes called "the mavericks of the solar system," have several remarkable similarities with planet earth. They contain considerable water. (About 38% of the mass of comet Temple 1 was frozen water.) Water is rare in the universe, but both common and concentrated on earth--often called "the water planet." Most of the remaining mass of a comet is dust, primarily the crystalline mineral olivine. Solid material that formed in space would not be crystalline. Olivine may be the most abundant of the almost 4,000 known minerals in the earth's crust and mantle. Asteroids and meteorites are similar in many ways to earth rocks. Surprisingly, a few meteorites contain salt crystals, liquid water, and living bacteria! Some asteroids have a chemical substance (kerogen) found in plants."[216]

---

214 Brown, *In the Beginning,* 121.

215 Brown, *In the Beginning,* 278-279.

216 Brown, In the Beginning, 116.

Wow! Finally we have solid evidence that life must have *evolved* elsewhere in the universe?! NOT!! We have evidence for a cataclysmic worldwide flood caused by the breaking up of "all the fountains of the great deep" and the shooting out into space of a volume of earthen debris from those fountains, exactly like God's Word describes.

It also caused torrential rains; but some of the water reached outer space where the temperature is minus 270°F, or absolute zero, before falling back to earth and instantly freezing all of those mammoths and other animals we find, who have remained frozen in some of the coldest climes like Siberia and Alaska.[217]

> "Some jetting water rose above the atmosphere, where it froze and then fell on various regions of the earth as huge masses of extremely cold, muddy "hail." That hail buried, suffocated, and froze many animals, including some mammoths."[218]

> "Fleshy remains of about 50 elephant-like animals called *mammoths*, and a few rhinoceroses, have been found frozen and buried in Siberia and Alaska. One mammoth still had identifiable food in its mouth and digestive tract. To reproduce this result, one would have to *suddenly* push a well-fed elephant (dead or alive) into a very large freezer that had somehow been precooled to *-150°F*. Anything less severe would result in the animal's internal heat and stomach acids destroying the food. If the animal remained alive for more than a few minutes, one would not expect to find food in its mouth."[219]

> [And keep in mind that this super frozen water was NOT heated up as it fell back to earth because, unlike our

---

217   (And no, that super frozen water would not have been heated up upon re-entry into the earth's atmosphere like space capsules or the Space Shuttle. These vehicles must orbit the earth at extremely high velocities (approx. 18,000 mph) so their centrifugal force will counteract the pull of earth's gravity, and therefore allowing them to stay in orbit. When they re-enter the earth's atmosphere at those velocities, the friction with the atmosphere causes extreme heat. Not so for water frozen to absolute zero that falls back to earth at a comparatively low speed and much less friction.)

218   Brown, *In the Beginning,* 121.

219   Brown, *In the Beginning,* 110.

space vehicles that hit the atmosphere at extraordinary speeds, the speeds necessary to keep them in orbit, and therefore the friction of their re-entry causes great heat, this super frozen water just fell back to earth at the relatively slow speed of any falling object in earth's atmosphere.]

There is only one plausible explanation for these instantly frozen creatures and that is the one given above. All others are pathetic attempts to explain away the obvious. Our academics have been taught to laugh at the idea of a worldwide flood (but in reality the laughs on them) so they just close their ears and their minds to all of the overwhelming evidence, write off all of those who believe otherwise as "stupid religious nuts" and continue to proudly parade around our colleges and universities, like the silly emperor with no clothes, as the silly professors with no sense. And unfortunately their condition is still hidden from the great majority of their students.

The waters rushing out from under the crust quickly eroded the sides of the ten-mile-high walls and spilled out across the earth, and this muddy deluge formed the sedimentary layers that cover the earth today, in the process sweeping up mass quantities of animal and plant life, an entire planets worth, and then rapidly burying and fossilizing them deep within those sediments. This is what we see today preserved in the "worldwide fossil record." (And if you'd like to get an idea of what this muddy flood spilling out across the earth looked like, we have just been given a glimpse in the recent Japanese tsunami, in March of 2011. Just look at all of those videos of the surging black waters that roared in from the ocean, filled with homes and barns and debris that marched across many miles of farm lands and towns and cities devouring everything in its path, and you can see a small-scale replica of what occurred 4,500 years ago on a worldwide scale!)

"Each side of the rupture was basically a 10-mile-high cliff. Compressive, vibrating loads greatly exceeded the rock's crushing strength in the bottom half of the cliff face, so the bottom half of the cliff continuously crumbled, collapsed, and spilled out into the jetting fountains. That removed support for the top half of the cliff, so it also fragmented and fell into the pulverizing supersonic flow. Consequently, the 46,000-mile-long rupture rapidly grew to an average width of about 800 miles all around the earth.

> "About 35% of the eroded sediments were from the basalt of the chamber floor. Sediments swept up in the escaping flood waters gave the water a thick, muddy consistency. These sediments settled out over the earth's surface in days, trapping and burying many plants and animals, beginning the process of forming the world's fossils."[220]

As the escaping water opened an ever-widening crevasse, the point was eventually reached where the opening became wide enough to substantially reduce the pressure on the earth's mantle directly below it. This caused the mantle there to buckle up which created the Mid-Oceanic Ridges. All other attempts at explaining the origin of these ridges in order not to contradict the evolutionary, old-earth model have proven somewhat fanciful, as well as failing to jibe with the existing scientific evidence. (Which is what scientific explanations are supposed to do, aren't they? As opposed to, say, supporting a particular group's strongly held evolutionary religious beliefs.)

> "Material within the earth is compressed by overlying rock. Rock's slight elasticity gives it springlike characteristics. The deeper the rock, the more weight above, so the more tightly compressed the "spring"—all the way down to the center of the earth.
>
> "The rupture path continuously widened during the flood phase... Eventually, the width was so great, and so much of the surface weight had been removed, that the compressed rock beneath the exposed floor of the subterranean chamber sprung upward."[221]

The currently accepted plate tectonics notion that "supposedly, material deep inside the earth is rising toward the crest of the entire Mid-Oceanic Ridge... [and then] moves laterally away from the ridge"[222] unfortunately defies the laws of physics and also is a poor explanation for the observed phenomenon; one that seeks to deny the more simple and much more obvious one.

---

220 Brown, *In the Beginning,* 121.

221 Brown, *In the Beginning,* 122.

222 Brown, *In the Beginning,* 108.

As the mantle buckled up along the Mid-Atlantic Ridge, gravity caused the now-separated continents to slide away on a cushion of the remaining "waters of the great deep."

> "As the Mid-Atlantic Ridge began to rise, creating slopes on either side, the granite **hydroplates** [continents or *tectonic plates* riding on water, hence *hydro*] started to slide downhill. This removed even more weight from what was to become the floor of the Atlantic Ocean. As weight was removed, the floor rose faster and the slopes increased, so the hydroplates accelerated, removing even more weight, etc. The entire Atlantic floor rapidly rose almost 10 miles. [It has, of course, since subsided.]

> "As the first segment of the Mid-Atlantic Ridge began to rise, it helped lift adjacent portions of the chamber floor just enough for them to become unstable and spring upward. This process continued all along the rupture path, forming the Mid-Oceanic Ridge. Also formed were fracture zones and the ridge's strange offsets at fracture zones. ... For a day or so, the sliding hydroplates were almost perfectly lubricated by water still escaping from beneath them."[223]

Also, these "fracture zones and strange offsets" all along the Mid-Oceanic Ridges cannot be sensibly or even intelligently explained by uniformitarian geologists.

As these continents travelled farther and farther apart the remaining water in the subterranean chambers spilled out, eventually covering the entire earth. "And the waters prevailed exceedingly upon the earth; and all the high hills that were under the whole heaven were covered." (Genesis 7:19) Then, as the lubricating waters under these sliding continents were exhausted, they ground to a "screeching" halt, buckling up all of the great mountain ranges of today:

> "Eventually, the hydroplates ran into resistance of two types. The first happened as the water lubricant beneath each sliding plate was depleted. The second occurred when a plate collided with something. As each massive hydroplate decelerated, it experienced a gigantic *compression event--* buckling, crushing, and thickening each plate. ... crashing

---

223 Brown, *In the Beginning,* 124.

hydroplates at the end of the continental-drift phase crushed and thickened each hydroplate for many minutes. Mountains were quickly squeezed up.

…"Naturally, the long axis of each buckled mountain was generally perpendicular to its hydroplates motion--that is, parallel to the portion of the Mid-Oceanic Ridge from which it slid. So, the Rocky Mountains, Appalachians, and Andes have a north-south orientation.

…"Friction at the base of skidding hydroplates and below sinking mountains generated immense heat, enough to melt rock, produce huge volumes of magma, and begin earth's volcanic activity. … Sometimes magma escaped to the earth's surface, producing volcanic activity and "floods" of lava outpourings, called *flood basalts*, as seen on the Pacific floor and the Columbia and Deccan plateaus.

…"As the new postflood continents rose out of the flood waters, water drained into newly opened ocean basins."[224]

And now we know the rest of the story, the true one that we were never taught in school. This is what actually happened to form the features of the earth that we see today. This is a scientific explanation, as opposed to the fairy tale pawned off as science that is still being taught to our school children today; not because of any preponderance of facts or evidence, but merely because it fits into the framework of a false, unscientific religion that long ago usurped the place of science, truth and reason. (Hats off to the father of lies for his accomplishment in getting *his* idiotic genesis account to its place of preeminence in our culture.)

Also, attempting to explain the features and dynamics of the earth's continents, volcanoes and earthquakes with *plate tectonics theory* is fraught with problems and scientific impossibilities. Earth scientists were right initially in advancing the idea of continental drift, but by refusing to acknowledge the earth's present features as the aftermath of a cataclysmic flood, they have been spinning their wheels ever since. And their attempts at explaining all that we see now, as a result of this refusal, have proven implausible. For one, the idea of *subduction*, of a huge continental plate slowly sliding underneath another plate over, what else, "hundreds of millions of years," is scientifically impossible according to the laws of physics. It would be akin to you trying

---

224   Brown, *In the Beginning*, 125.

to shove your flattened stiffened hand and fingers slowly through your skull. Even if you could stand the pain you wouldn't get subduction, you would get destruction.

> "Pressure inside the earth increases with depth. So, if one tried to depress a plate 30 miles or more below another plate, tremendous upward pressure from below would quickly prevent that much depression. Consequently, subduction—necessary for plate tectonics—could not begin, even if the plate were colder and, therefore, denser."[225]

Now let's leave the study of *how* that global flood occurred and look further into the worldwide compelling evidence *for* that same deluge. Let's begin with the sedimentary layers that cover the earth:

> "Sedimentary rocks are distinguished by sharply defined layers called **strata**. Fossils almost always lie within such layers. Fossils and strata, seen globally, have many unusual characteristics. A little-known and poorly-understood phenomenon called **liquefaction** explains these characteristics. It also explains why we do not see fossils and strata forming on a large scale today.
>
> ...*"Liquefaction--associated with quicksand, earthquakes, and wave action--played a major role in rapidly sorting sediments, plants, and animals during the flood. Indeed, the worldwide presence of sorted fossils and sedimentary layers shows that a gigantic global flood occurred. Massive liquefaction also left other diagnostic features such as cross-bedded sandstone, plumes, mounds, and fossilized footprints."*[226]

> "Sedimentary rocks need water for cementing the tiny particles of sediment together. Where did the water come from? Evolutionists resolve this dilemma by waving the magic wand of "time." They insist that the sediments are washed into rivers and lakes and oceans and then fall

---

225 Brown, *In the Beginning*, 159.

226 Brown, *In the Beginning*, 169. (For a more in-depth study of liquefaction and the sedimentary layers that cover this earth, see pages 169-181, *Liquefaction: The Origin of Strata and Layered Fossils* in Dr. Brown's book.)

to the bottom and pile up in layers. After many years, the sediments harden in layers, with those on the bottom hardening first. The problem with this theory is that there are beds of sedimentary rock everywhere on top of the earth's crust. Most of the large mountain chains are made of sedimentary rocks. How did all those sediments collect and become cemented together to make towering mountains? The logical answer is a worldwide Flood."[227]

The problem also is that these sedimentary layers not only cover the entire earth and are a mile thick in many places, but they also cover thousands of square miles where each successive layer is dropped onto the one below it with absolutely <u>no</u> sign of water or wind erosion, chemical deterioration, plant roots or animal burrowing, or the passage of any time whatsoever. Look at the walls of the Grand Canyon and you will see that these lines in between each layer, going for mile after mile, could have been drawn with a laser! Plant rooting, animal burrowing and wind and rain erosion over hundreds of millions of years would produce tremendous signs of the passage of time between each successive layer. (And therefore, for one, the lines on the side of the Grand Canyon would be zigzagged and jagged and discontinuous in places.) But there is none whatsoever. Where is the evidence of *any* passage of time, Mr. Uniformitarian evolutionist, much less "hundreds of millions" of years?! So much for the Darwinists obviously erroneous and unsupportable claim that these sedimentary layers all across this earth were laid down over hundreds of millions of years. There is no evidence to support their claims.[228]

"The earth's sedimentary layers are typically parallel to adjacent layers. Such uniform layers are seen, for example, in the Grand Canyon and in road cuts in mountainous terrain. Had these parallel layers been deposited slowly over thousands of years, erosion would have cut many channels in the topmost layers. Their later burial by other sediments would produce nonparallel patterns. Because parallel layers are the general rule, and the earth's surface erodes rapidly, one can conclude that almost all sedimentary layers were deposited rapidly relative to the local erosion rate—not over long periods of time."[229]

---

227 DeRosa, *Evidence for Creation*, 35-36.

228 Morris, *The Young Earth*, 102.

229 Brown, *In the Beginning*, 11.

The sedimentary rock layers that cover this earth are incontrovertible evidence for a global flood, and for nothing else. There is no other intelligent interpretation of the data.

> "Rock formation can be explained by means of two processes, compaction and cementation, both of which could have occurred within one year after the waters of the Flood began to subside.
>
> "Global floodwater would have collected a huge amount of sediment. When the flood water came to rest, the sediment would have begun to fall out. At the beginning… there would have been a great deal of pressure on the bottom layers. This would cause compaction.
>
> …"The second step is cementation. The glue that hardens the rock is generated by the warm water and minerals produced by the effects of the Flood. It does not take a long time for the rock to harden into layers. … Remember, making sedimentary rock is like preparing cement."[230]

And cement can become as hard as rock in as little as a few hours.

The Grand Canyon[231] is one of the world's most striking testimonies to the worldwide flood. The idea that a measly river carved this canyon and all of its side canyons through a *raised plateau* by defying gravity is a testament both to man's stupidity and to the level of willful ignorance that he can rise to if so driven by tightly-held, false-religious beliefs: "There is no God. There was no worldwide biblical Flood. Evolution is true. And that is the end of the discussion." And science, common sense, and all the evidence be damned.

Also, all across this world there are countless examples of these layers of sedimentary rock being turned, twisted, compacted, accordioned (folded back and forth upon them self like the folds of that musical instrument) and tilted. The Grand Canyon has many such places. Only a worldwide catastrophic aqueous event could have produced this. And this evidence is absolutely damning for uniformitarian geology, which is blind to the signs of this catastrophe written all across the earth. There are innumerable instances

---

230 DeRosa, *Evidence for Creation*, 36.

231 (For an excellent in-depth study on what really made the Grand Canyon and all of the features associated with it, see pages 183-219, "The Origin of the Grand Canyon" in Dr. Brown's book, *In the Beginning*.)

around the world where we can see that a number of these sedimentary layers have together been squeezed, bent, folded, curved and compacted into wave formations (as well as accordioned) which could only have occurred when these sediments were still *in a putty-like state*. Yet supposedly, according to evolutionists and uniformitarian geologists, these layers were laid down over hundreds of millions of years, with millions of years separating the beginning of the deposition of each successive layer on top of the one beneath. Which means that, according to them, <u>all</u> of the layers would have long-since been hardened into solid <u>rock</u>, not <u>putty</u>. And this *putty-like state* can only occur for a very short period of time, measurable in days or weeks, immediately after the sediments were deposited and before they had a chance to completely harden. The sedimentary layers therefore had to have been laid down very rapidly one upon the other up to thousands of feet thick and covering hundreds of thousands of square miles and then very soon thereafter, before they had time to harden and cement themselves, were deformed and bent, folded and curved, squeezed into wave and accordion formations due to the continents "skidding to a stop". And of course this is exactly what did occur as the sediments were rapidly laid down by the muddy flood waters, separated into distinct layers by the process of liquefaction, began the process of cementing, and then as the sliding continents came to a halt they were compressed, bent, folded and buckled up all over the earth. The explanation of hundreds of millions of years for the deposition of these layers is absurd, and was only thought up to support the theory of evolution by aiding it with preposterous spans of time where "anything is possible."

In addition to the Grand Canyon there are countless other areas where these sediments had to have been deformed while still soft and unconsolidated. The Appalachian and the Rocky Mountains have many such incidences. The deformation of soft sediments all across this earth are undeniable evidence of a catastrophic flood.[232]

Hardened rock is brittle, and it doesn't squeeze, bend, fold, curve or compact into wave formations. It breaks! It cracks! It busts up! If these sedimentary layers were sifted down over hundreds of millions of years as Darwinists claim, then how could all of those layers still have been in a putty-like state, millions upon millions of years later, in order to have been able to be squeezed and bent and folded and curved and compacted into wave formations <u>before</u> they hardened and became brittle? Answer: of course they

---

232 Morris, *The Young Earth,* 112-113.

were not sifted down over hundreds of millions of years. They were laid down rapidly by a global flood.

Before the time of Charles Lyell[233] scientists used their common sense and saw the features of the earth's surface—massive and deep sedimentation, the Grand Canyon, the white cliffs of Dover, seashells on the tops of every mountain range, etc.—and logically and naturally concluded that they were the result of the Great Flood. But that was before logic, rationality and science took a back seat to the religion of Darwinism, when God and His Bible had to go, and anything attributed to God or His Word (like a world-wide, biblical flood) had to go as well.

And the evidence is overwhelming… All over the world there are sites of mass burials, in caves and crevasse's, which are crammed with countless fossils. Many times these animals originated from completely different climactic areas, and they are strewn about in a jumbled mass grave. The only way this can be explained is due to an even much more massive worldwide flood.

Millions of large animals and mammoths are found in Alaska and Siberia and many are completely preserved as they were right before they instantly expired with their hair and their flesh unspoiled, sometimes either standing or kneeling with food in their mouths. Also very well preserved are their red blood cells and eyes, as well as only a partial separation of the cellular water. This is the result of an abrupt and continuous deep freeze that occurred, as we said before, as the split in the earth's crust widened and the escaping waters of the great deep were under less pressure and hence slowed to the point where they could no longer escape earth's gravity, then froze in space to absolute zero, and fell back gently at comparatively low speed (compared to meteors and space capsules) while therefore maintaining their bitter, far below zero, temperature.[234]

---

233   (Lyell, an atheist determined to undermine the Bible, was a geologist in the 1800's who championed the idea of uniformitarianism (slow, uniform changes in the earth's features over great periods of time), and rejected the biblical concept of catastrophism as well as the obvious effects on this planet of a global flood. The idea of uniformitarianism also fit in neatly with Charles Darwin's new theory of evolution (indeed Darwin had a copy of Lyell's book on his famous Beagle voyage) which soon became all the rage with the European intellectuals of the day, and which demanded the outrageous lengths of time—provided conveniently by Lyell and his unscientific interpretation of geologic features—to give Darwin's theory a better shot at believability.)

234   Huse, *The Collapse of Evolution*, 47-48.

Of course, we know for a scientific fact that it is impossible for all of these cellular tissues to have been preserved for hundreds of millions of years. But this fact is ignored so the Big Lie can remain, and can continue to be taught.

Petrified logs are found by the thousands throughout the fossil record, with their cellular structures intact by silica replacement, but yet none are forming today. They have been stripped of their branches and their bark is still there which shows a very sudden and deep burial occurred before any rotting could happen. Obviously these trees, indeed whole forests, were pulled up by an aqueous event of cataclysmic energy. They were dragged along by this enormous flood and then laid down and deeply buried in their current locale.

Another spoke in the wheel of the completely implausible uniformitarian explanation of things are what are called *Polystratic trees*. These fossilized trees are found throughout the world extending up through layers of sediment. Yet these sediments were supposedly laid down slowly and uniformly over millions of years??!! These trees could not have fossilized. They would of course have rotted.[235]

Science is ruled by observation. That is the very method that gives rise to science itself. How can anyone in his or her right mind look at all of these Polystratic trees, trees that stand upright through "millions" of years of sediment and are found all over the globe, as well as animals, including a fossilized whale in California standing on its tail through "millions of years" of sediment, and conclude anything other than those trees and those animals and that whale were placed there when the sediments were, and therefore both were deposited rapidly, and therefore the ridiculous ages for these sediments that the lie of evolution demands are patently absurd? Indeed, the mental gymnastics one has to perform in order to conclude otherwise are truly amazing. So they just ignore it. Unfortunately, people do believe what they are taught. No matter how lame. And they keep on teaching it to generation after generation. And the curse goes on and on…

And then there are those markings, called ephemeral, which are found copiously throughout the fossil record. The imprints of rain, tracks of animals, ripple marks and even worm trails. None of these are being formed today. Because again, they could only have been preserved by the rapid burial right after they were made, by new sediments.

---

235   Huse, The Collapse of Evolution, 51-52.

And then there are the many preserved soft tissues within extremely old sedimentary layers. The idea that these were slowly laid down on top of these organisms, without them rotting immediately, as well as not rotting away over millions and millions of years, is really quite silly. They had to have been buried recently and deeply in order for us to find their soft tissues still intact.[236]

And in summary… The fossil record clearly shows us by such things as Polystratic trees, petrified logs, soft tissues, woolly mammoths, huge graveyards of fossils, ephemeral marks, etc. etc. that their cause is a massive and cataclysmic worldwide flood. And the game is over and the score is in… The Genesis Flood of the Bible, 7 trillion! The mythological joke of *uniformitarianism*, 0. Zip. Nada. Bupkis. Zilch.[237]

And this would all be common knowledge in our universities, colleges, elementary and high schools and throughout our country, if it weren't for the fact that the lies of the evil one cover this earth like a thick sediment, and people believe whatever they are taught.

"You who laid the foundations of the earth, So that it should not be moved forever, you covered it with the deep as with a garment; The waters stood above the mountains. At your rebuke they fled; At the voice of your thunder they hastened away. They went up over the mountains; They went down into the valleys, To the place which you founded for them. You have set a boundary that they may not pass over, That they may not return to cover the earth." (Psalms 104:5-9)

Again, as we stated in chapter 6, "The Word of God" (in the larger work "All the Secrets") …The Bible says that there was a catastrophic worldwide flood that occurred about 4,500 years ago. Faux scientists say that all of the features found on the earth today, including the Grand Canyon, can be explained by the slow processes of upheaval and erosion continuing slowly over millions of years. Again, the Biblical account is scientifically accurate. The evidence for a worldwide flood is overwhelming and incontrovertible; yet it is blackballed by the baby-mutilating Democrat evolution indoctrinates who possess a stranglehold over our media, entertainment industry and educational system. There are trillions of recently deposited seashells on the tops of the highest mountains. Coal and oil formations demand the rapid and very deep burial of an extraordinary volume of living matter. The fossil

---

236 Huse, *The Collapse of Evolution*, 52-54.

237 Huse, *The Collapse of Evolution*, 54.

record which covers this earth could only have been caused by the rapid worldwide burial of trillions upon trillions of living things (fossils can <u>only</u> form under conditions of rapid and <u>deep</u> burial- a dead fish lying on an ocean floor, or a dead elephant lying on the ground have no chance whatsoever of being fossilized). It is nothing more than an unrealistic and unscientific fantasy to think that the Grand Canyon and all of her side canyons were formed by a river slowly eroding a *raised plateau* over any period of time. All the scientific and erosion rate and sedimentary evidence call for the Canyons rapid formation by the catastrophic runoff of a massive amount of water over a very short period of time. On a worldwide scale sedimentary layers thousands of feet deep, with absolutely no evidence of time passage or erosion between each successive layer (their transition surfaces are perfectly smooth and flat), and extending in many places uniformly for hundreds of thousands of square miles cry out for only one possible scientific explanation, that of a catastrophic worldwide flood! And we could go on. But once again the Bible proves itself scientifically and historically accurate.

"Then the Lord saw that the wickedness of man was great in the earth, and that every intent of the thoughts of his heart was only evil continually. And it sorrowed the Lord that he had made man on the earth, and it grieved him at his heart. And the Lord said, I will destroy man whom I have created from the face of the earth." (Genesis 6:5-7)

Approximately 4,500 years ago God allowed the waters under the crust of the earth, the "fountains of the great deep," to erupt and flood the entire planet. He did this because of the wickedness of man. But how can anyone think that the evil of this present generation is any less than theirs? And in this country God has shed his grace on us in ways never even dreamt of by those people thousands of years ago, and yet how do we repay his kindness and mercy and abundance? For starters, we shove the bald-faced, scientifically-absurd, Beelzebub-inspired lies of evolution and the big explosion down the throats of our school children; and then we top that off with mutilating millions of his babies to death each year, and going out and foolishly voting for the same. If we look across this earth with open eyes we can see the aftermath of his last judgment. The next one will not be by water, but it will be no less severe. God's approaching final judgment of wicked, unrepentant man will be by fire. (See chapter 9 "Bible Prophecy".)

But as noted in chapter 7 ("Jesus of Nazareth and Gods free gift of eternal life" in the larger work) , he has made a way of escape for us. Indeed the history of Noah's ark and the worldwide deluge are themselves a prophecy of Christ and His work of redemption. For as the ark saved righteous Noah and his family from destruction in the flood, so Jesus saves all who place their trust and faith in Him and His righteousness, from the horror of eternal separation from God and all that is good in a place of everlasting torment. Our school children are in desperate need of hearing this truth. Who has the guts to tell them? O'Reilly? Tucker?? Fox news? Anybody there?

Human "Races"

Question: *OK then, the earth is young just like the Bible and all of the scientific evidence says, and all 6 billion of us alive today are descended from Noah and his wife and children who walked off of the ark roughly 4,500 years ago. But what about all of the races of man? Eskimos, and Caucasians, and Negroes, and Australian Aborigines, and Chinese, and Japanese, and Polynesians, and American Indians, and India Indians; tell us where all of these races came from if they didn't have millions of years to "evolve?"*

Answer: We don't have to tell you. God's Word already has. It's in Genesis 11; in the story of the Tower of Babel…

"Now the whole earth was of one language and of one speech. And it came to pass, as they journeyed from the east, that they found a plain in the land of Shinar, and they dwelt there. And they said to one another, 'Come, let us make bricks and bake them thoroughly.' And they had brick for stone, and they had asphalt for mortar. And they said, 'Come, let us build ourselves a city, and a tower whose top may reach unto heaven; let us make a name for ourselves, lest we be scattered abroad over the face of the whole earth.' But the Lord came down to see the city and the tower which the children of men had built. And the Lord said, 'Behold, the people are one and they all have one language, and this is what they begin to do; and now nothing that they propose to do will be restrained from them. Come, let <u>Us</u> go down and there confuse their language, that they may not understand one another's speech.' So the Lord scattered them abroad from there over the face of all the earth, and they stopped building the city. Therefore its name is called Babel, because the Lord confused the language of all the earth; and from there the Lord did scatter them abroad over the face of all the earth." (Genesis 11:1-9)

[A side note, look at "Come, let Us go down…" That is another reference to the triune nature of the One God.]

This "confusion of the languages" was also the beginning of the differentiation of the distinct "racial" types that we see today across this globe. The reason "racial" is in quotes is because there really is only one race, the human race, and all the differences we see are merely the result of "variations within a kind" that we discussed earlier. And whether these variations in skin color, body size and body type, and distinct facial features were brought about by the Lord at the same time that he confused the languages, or whether they developed over successive generations after each language separated each group from one another, is a question that remains. But the fact that that is where the separation of "racial" types occurred, or began to occur, is fairly obvious.

For one, variations in skin color are something that is brought about merely by varying within the epidermal cells the amount of the skin-coloring pigment *melanin*…

> "We all have the same coloring pigment in our skin—melanin. This is a dark-brownish pigment that is produced in different amounts in special cells in our skin. If we had *none* (as do people called albinos, who inherit a mutation-caused defect, and cannot produce melanin), then we would have a very white or pink skin coloring. If we produced a little melanin, we would be European white. If our skin produced a great deal of melanin, we would be a very dark black. And in between, of course, are all shades of brown. There are no other significant skin pigments.
>
> …"This situation is true not only for skin color. Generally, whatever feature we may look at, no people group has anything that is essentially different from that possessed by any other. For example, the Asian, or almond, eye differs from a typical Caucasian eye in having more fat around them. Both Asian and Caucasian eyes have fat— the latter simply have less.
>
> "What does melanin do? It protects the skin against damage by ultraviolet light from the sun. If you have too little melanin in a very sunny environment, you will easily suffer sunburn and skin cancer. If you have a great deal of melanin, and you live in a country where there is little sunshine, it will be harder for you to get enough vitamin D (which needs sunshine for its production in your body). You may then suffer from vitamin D deficiency, which could cause a bone disorder such as rickets."[238]

238　(From article, *Where did the human races come from?* @ christiananswers.net)

So the need for vitamin D alone could have caused the divergence, by the natural selection of those humans better suited for their sunny or cold environments, of the various skin colors we find in the different "races" such as European Caucasian and African Negro. For these continents obviously have entirely different climates, average temperatures and levels of sunshine. Or God could have allowed for those different needs by altering skin color and other features when he confused the languages, knowing in advance where these separate language-groups would end up in their migrations after He "scattered them abroad over the face of all the earth." Either way though, we can dispose of the false anthropological notion that because there are obvious differences between racial groups and sub-groups, there had to be vast periods of time for these differences to be brought about. That is a false notion that only survives because of a rejection of the accuracy of God's Word concerning this matter, and an acceptance of ridiculous time periods for life on earth in order to support the lie of evolution.

PART III

THE RELIGION OF NATURALISM

"For they exchanged the truth of God for a lie, and worshipped and served the creature more than the Creator…" (Romans 1:25)

It should be obvious now to any reader who seeks the truth that the theory of evolution is not only a joke, and a lie, but it is also not science. It is of course a religion. No mere theory this scientifically-impossible, this bereft of any credible evidence whatsoever and indeed this irrational and idiotic could have survived within the confines of science for even five minutes much less have thrived for over 100 years. No, there has to be something going on here much more powerful than just a false theory. There has to be a deep religious fervor involved in order to drive such a colossal lie as this into a position of total dominance in the classrooms and the cultures of an entire world. It is no different than the reason why the satanic, genital-mutilating, Jew and Israeli hating religion of Islam has been driven into a position of total dominance throughout the Muslim world. Blind, irrational, intolerant religious fervor; supported also by the fear of academic ridicule and exile on the one hand, or death on the other, for anyone who would dare to challenge either religions' beliefs.

"Evolution dismisses intelligence and assumes random chance to be the mechanism responsible for material reality—a physical presupposition that is not drawn from empirical evidence. Evolution, therefore, is an irrational belief. It assumes randomness and chance, not design, to be the governing principle of reality, a view, again, without hard empirical data to support it."[239]

---

239   DeRosa, *Evidence for Creation*, 16-17.

"A view, again," that is also logically and scientifically absurd.

Satan is the god of this world, and the father of lies. His lies cover this earth like a blanket, and people believe whatever they are taught. He hates the gospel of Jesus Christ and will do anything in his power to keep as many people as possible from hearing it and believing in it. One of his greatest achievements has been in using science, indeed *usurping* it, to serve his diabolical goal. The gospel of Jesus Christ is intimately tied to the integrity of the Bible. So Satan, the great deceiver, along with his intellectual idiots, has used every means he can to lie about the Bible, to vilify it, to falsely demean and disparage it, in order to discredit it in the eyes of a gullible and uneducated world. And the integrity of the Bible is intimately tied to the truth of the Book of Genesis. Indeed Genesis is the *very foundation* of God's Word. Look at the very first verse in Genesis, "In the beginning God created the heaven and the earth." (Genesis 1:1) Satan took aim at this verse with the theory of evolution (as well as with that of the big explosion) and his dupes have been continuing the assault ever since. Destroy the foundation of God's Word through the lies of faux science and you have destroyed the integrity of the Bible in the eyes of this foolish and deceived world. And you have also cut off the same world from their only hope of salvation, the glorious gospel of Jesus Christ. Their only hope of avoiding an eternity of absolute horror burning alive forever and ever with the Liar and his demons in the Lake of Fire (see the Book of Revelation). This is what Satan has achieved through the widespread acceptance of his religion of Naturalism and its twin theories of the absurd.

Two worldviews are in conflict here. The creationist worldview is based on God's Word, has faith in a Creator-Intelligence outside the universe, believes in Absolutes in morality and truth, and that man is accountable to God. The evolutionist worldview is based on man's word, has faith in unknown internal processes within the universe, believes in relativism in morality and truth, and that we are accountable to ourselves only.[240] The creationist worldview is supported by all of the facts and evidence, the laws of science, and common sense. The evolutionist worldview is scientifically absurd and built on a foundation of lies.

Instead of a search for truth which has always defined science, a new insidious approach has taken root. It is actually the search for lies, as long as those lies fit their religion. This religion, the religion of naturalism, controls

---

240 DeRosa, *Evidence for Creation*, 14.

science and says that no explanation outside of the universe, no "super" natural explanation, can even be considered. Which is absolutely moronic because, as we have clearly seen in this book, the only way for the universe and for life on earth to exist is obviously due to a supernatural explanation, one not tied to nor limited by the natural world. This approach is no less ignorant than if someone said we are going to search for the origins of your vehicle, but we can't look anywhere outside of your vehicle. No Detroit, or Japan, or anywhere else cars are made. We can only look INSIDE of the vehicle. That moron would of course be laughed off the face of the earth. But with the religion of naturalism they get away with it because of their complete and total control over our educational system, our entertainment industry and our baby-killing, Democrat-controlled media.[241]

> The biblical statement, "the Spirit is truth" (1John 5:6), is saying that truth exists outside of the confines of this physical universe; that truth is transcendent from the material world. It was here before the world began; and before man and his lofty intelligence had their beginnings. And if one seeks the truth, it is there where he must find it. To do otherwise—to seek the truth within the confines of this universe of matter alone—is to seek it in vain. It is to look for it, like love, "in all the wrong places." This is the source of one of the major errors in science today, which has turned certain disciplines within it into more of a primitive *religion*; and one that holds sway over far too many *scientists*. And that is the religion of *Naturalism*, which declares that there is no *meta*-physical truth or reality—meaning that there is no truth or reality that exists *above* or—outside of the physical world. So all truth and all reality must be found within and only within the confines of the physical world. Including the *origins* of this physical world.[242]

But let's not be too hard on our evolutionist friends. (I know. I know. A little too late for that.) They believe it because it is the only thing they have ever been taught.[243] They have been lied to throughout their journey through our educational /indoctrination system by teachers/authority figures who

---

241 Morris, *The Young Earth*, 6.

242 (From the chapter "Truth" in the larger work, "All the Secrets.")

243 (Again, the lies of Satan, the great deceiver, cover this earth like a blanket. And people believe whatever they are taught.)

were also proselytized, people who they of course looked up to and respected. And their indoctrination included being taught to laugh at and mock stupid "flat earthers" who believe in a Creator. Essentially cutting them off, totally, from any chance of ever hearing, and thus learning, the truth. They accept without a critical thought the sole "legitimacy" of their view and so the lie is passed down, passionately, to successive generations. Just as in the case of any of the world's many, many other false religions[244]

"For I, the Lord your God am a jealous God, visiting the iniquity of the fathers upon the children unto the third and fourth generation of those who hate me" (Exodus 20:5) …and of those who teach that His Genesis is a lie.

> "The existence of God is nowhere defended by Scripture. This fact is taken as being obvious. …Neither is there any doubt as to His sovereign authority over His creation or what our attitude should be toward Him as Creator. He has the right to set the rules. We have the responsibility *to obey and rejoice in His goodness*, or disobey and suffer His judgment.
>
> …"Those who oppose the Creator are opposing the One who is the absolute authority—the One who sets the rules and *keeps them*.
>
> "In the Book of Judges it is stated: "In those days there was no king in Israel; but every man did that which was right in his own eyes" (Judg. 17:6). People today are little different. They want evolution taught as fact and the belief in creation banished because they, too, want to be a law unto themselves. They want to maintain the rebellious nature they have inherited from Adam, and they will *not* accept the authority of the One who, as Creator and law-giver, has the right to tell them exactly what to do. This really is what the creation/evolution conflict is all about."[245]

In his book, *The Devils Delusion*, David Berlinski describes one particularly fanatical zealot, Hector Avalos, "a professor of religious studies at Iowa State University, and an avowed atheist …[who is] much occupied in denouncing theories of intelligent design…"

---

244 Morris, *The Young Earth*, 6.

245 Ken Ham, *The LIE: Evolution* (Green Forest, AR: Master Books, 2007), 80-81.

"He is a member in good standing of the worldwide fraternity of academics who are professionally occupied in sniffing the underwear of their colleagues for signs of ideological deviance."[246]

Still other academics, which have little inclination for the sniffing, are still resigned to accept their situation quietly, meekly and yet safely, as demonstrated by this comment from a colleague of Berlinsky's… "'Darwin?' a Nobel laureate in biology once remarked to me over his bifocals. 'That's just the party line.'"[247] And Ben Stein, in the movie *Expelled: No Intelligence Allowed*, had a field day with these underwear-sniffing, modern-day McCarthyites and their relentless witch-hunt of those scientists and professors who dare stray from the evolutionary party line. ("Sieg Heil!") Of course the reason it's the party line, and you must stick to it or else, is that, unlike scientists, religious fanatics have no interest in entertaining conflicting views or evidence. They only seek to defend their false doctrines from any exposure, and at all cost.

"Teachers, professors, and educators are gagged and told not to utter the words **God, Creator, Creation** and even **Intelligent Design.** When such words are spoken, or mere inferences are posed about alternatives to evolution in a classroom, sirens and alarms go off, alerting "gravediggers" like the ACLU, the National Academy of Sciences, the American Association for the Advancement of Science, and many others. Armed with their legal clubs and shovels, they rush to silence such utterances, hoping to bury them forever. If this sounds like something out of a police state, you're right. Unfortunately, this is happening in America." [In *Nazi, Demonrat* America.]

"The goal of the self-appointed censors mentioned above is to turn the U.S. into a secular state, where belief in God is outlawed from the public square and denied all government support. One of their tactics is to suppress all truth about the existence of a Creator in connection with science in our nation's schools. God is to be completely censored out and replaced with naturalistic evolution. These "elites" use the fallacious cry of "separation of

---

246 Berlinski, *The Devil's Delusion,* 52.

247 Berlinski, *The Devil's Delusion,* 192.

church and state"-- which cannot be found anywhere in our Constitution, but is actually part of the former Russian Constitution (1936)-- used to silence all critics."[248]

Dosvidanya, Comrades. The only difference between the totalitarian Democrat-Left in America and the leaders of outright communist countries like North Korea, Cuba, China, Venezuela and California, is that here their power is still limited by our Constitution. But at the rate they are shredding it, it is only a matter of time before they get that obstacle out of their way as well.

But there is hope. Because just as Genesis is the foundation of God's Word, from which we can know the truth of Christianity and the saving grace of Jesus Christ; so also is Satan's genesis, the theory of evolution, the foundation of his godless, lying, baby-mutilating, false-accusing, America-hating, filthy, iniquitous Democrat-Left. Destroy their theory and you destroy their foundation. You tear the rug out from under them. There are so many conservative and Christian commentators and authors and pundits and leaders and preachers and teachers and politicians who are fully engaged in the war against the Democrap/Left, but they are unfortunately clueless when it comes to their most effective weapon of all. And that is shooting the deadly arrow of actual scientific truth through the heart of Satan's genesis. They cannot do this because they believe the lie as well. Sadly, as much as they know about the truth on so many important subjects like proper education; our national defense; the role of the judiciary; insane left-wing judges and their incredibly insane, unconstitutional rulings; the war against Islamic lunatics; our precious freedoms and our economic prosperity; the Democrat's terrible America-destroying agenda;[249] they don't know the truth about the lie that props it all up. They fight the war in vain. They flail away, blindly, beating nothing but the air, while Satan, the child-sacrificer of old, the father of the baby-killing Democrat/Left laughs his ass off.

In New Guinea, in...

    ... "one cannibal tribe ...men would race into a village, grab a man by the hair, pull him back, tense his abdominal muscles, use a bamboo knife to slit open his abdomen, pull out his intestines, cut up his fingers, and

---

248   (Tom De Rosa, *Creations Studies Institute,* in a March 2009 Newsletter)

249   (See chapters 2 and 13 in the larger work, "All the Secrets of the Universe.")

while he was still alive, eat him until he died. People hear that and say, "Oh, what primitive savages!" They are not "primitive" savages; their ancestor was a man called Noah. ... Noah had the knowledge of God and could build ships. His ancestors could make musical instruments and they practiced agriculture. What happened to those New Guinea natives is that, somewhere in history (as Romans 1 tells us), they rejected the knowledge of God and His laws. And God turned them over to foolish, perverse, and degenerate things.

[And in the same way God has turned over the satanic, child-sacrificing, lying, baby-mutilating, racist black-baby-exterminating, false accusing Democrats who also control our media, educational system and entertainment industry, over to what is called a reprobate mind. A mind that is literally insane. In case anyone is wondering how the entire Democrat Party and media could embrace a total, certified moron like Ocrazyo Cortez, or how one of their "esteemed" Democrat-jackass congressmen could actually, seriously be concerned about the island of Guam "tipping over."]

"However, this same degeneracy (this same rejection of God's laws) can be seen in so-called civilized nations that cut people up alive all year long (one and a half million of them in the United States each year), and it is legalized. This is what abortion is—cutting up people alive and sucking out the bits and pieces. The so-called "primitive tribes" had ancestors who once knew the true God and His laws. As they rejected the true God of creation, their culture degenerated in every area. The more our so-called "civilized nations" reject the God of creation, the more they will degenerate to a "primitive culture.""[250]

And this is the same thing that is happening in our nation; the Democrat-Left and their father's theory of evolution are "progressing" us into a primitive culture that is controlled by intellectual savages. They too have "rejected the knowledge of God and His laws." They deny Him as Creator, and instead worship someone they can more easily relate to, a clueless imbecile, Random Accidental Chance. ...What say you, O'Reilly? Tucker? Greg? Tyrus? Still believe in evolution?

---

250 Ham, *The LIE: Evolution,* 84.

Darwin's Deadly Fruit

"For a good tree does not bring forth corrupt fruit, neither does a corrupt tree bring forth good fruit. For every tree is known by its own fruit. For men do not gather figs from thorns, nor do they gather grapes from a bramble bush." (Luke 6:43-44)

The fruit of Darwin's godless and unscientific theory has poisoned this world in the last century in ways that he could never have envisioned, and certainly never intended. But lies have consequences. The hidden demonic nature and evil intent of the theory of evolution can not only be seen in its attack on the truth of Genesis, Christianity, the gospel of Jesus Christ and science itself; and on its destruction of the moral basis of our society by helping to create and then empower the liberal, "progressive" (regressive), satanic, child-sacrificing, racist, black-baby exterminating, "transgender"-child-mutilating, false-accusing Democrat-Left; but it is also made evident in a number of cases of mass genocide in the twentieth century that it bares direct responsibility for.

Most people do not know that the full title of Darwin's book ends with this prophetic little phrase, "the preservation of favored races in the *struggle* for life." And, oh my, what a toll that struggle has exacted on the **un**-favored human races in the last hundred years.

> "Darwin's theory is an idea that wicked men have used as a rationale for mass murder; it is also an instrument of spiritual death for millions who embrace its lie."[251]

> "The two most notorious and blood-soaked political movements of the twentieth century, Nazism and Communism, both rejected God and were animated by the idea of evolution.

> "It was Darwin's theory—carried to its logical conclusion—that led to the death of some 11 million people at the hands of German Nazis. Darwin's notebook jottings became fodder for Hitler's guns. The German leader was a devout evolutionist. [And not even remotely a "Christian" like the historical revisionists on the Democrat-Left like to claim.] He was determined to create a super race by eliminating so-called inferior races. Hitler wrote in *Mein*

---

251 Tom DeRosa, *Evolution's Fatal Fruit: How Darwin's Tree of Life Brought Death to Millions* (Fort Lauderdale: Coral Ridge Ministries, 2006), 190.

Kampf: "The stronger must dominate and not mate with the weaker, which would signify the sacrifice of its own higher nature."

"Hitler tried to speed up evolution—to help it along. "The German Fuhrer, as I have consistently maintained, is an evolutionist," British evolutionist Sir Arthur Keith wrote in the 1940s. "He has consciously sought to make the practice of Germany conform to the theory of evolution." And millions suffered and died in unspeakable manners because of it.

"Likewise, Karl Marx, the founder of communism, found in evolution exactly what he needed: a pseudo-scientific foundation for his godless worldview. So too did Vladimir Lenin and communist henchman Josef Stalin.

"Combined, these communist leaders and others killed more people than all those killed in all religious wars. The "rough estimate" from the *Black book of Communism* is that the "total approaches 100 million people killed." Stalin, Mao, Pol Pot, and all the rest are the greatest mass murderers of all time—all complements of evolution."[252]

"There are those, usually at the universities, that have zealously dogmatized the world with the canons of evolutionary thinking. They have become the priests in their own religion and work with an evangelistic energy to spread the faith of evolution. **Yet this lie has had terrible practical consequences—among them abortion, Nazism, communism and different forms of socialism.**"[253] [Emphasis mine.]

…Hmm. Like the *different form of socialism* being shoved down the throats of Americans today (2009) by the Islamic Obama administration and his Democrat congress.

"In the limitation of this living space lies the compulsion for the struggle for survival, and the struggle for survival, in turn, contains the precondition for evolution." (Adolf Hitler)

---

252   (From *Impact*, November 2009, a circular from Coral Ridge Ministries)

253   DeRosa, *Evidence for Creation*, 69-70.

And then we have that visionary hero of so many American university professors, Karl Marx, who wrote this in a letter to Ferdinand Lassalle:

> "Darwin's book is very important, and serves me as a basis in natural science for the class struggle."[254]

Karl's *class struggle* (a variation of which our hapless president Obama is trying to orchestrate today—April 2012—for his reelection purposes of course) ultimately resulted in the slaughter of a hundred million human beings.

In addition to being the ideological inspiration for some of the world's greatest genocidal maniacs, Darwin's theory also has on its resume being the rationale behind modern-day racism:

> "Arguments for racism may have been common before 1859, but they increased by orders of magnitude following the acceptance of evolutionary theory."[255] So says leading <u>evolutionist</u>, Stephen Jay Gould.

In his book, *The Descent of Man*, Darwin himself had this to say, "at some future period, not very distant as measured by centuries, the civilized races of man will almost certainly exterminate, and replace, the savage races throughout the world." Those "savage races" he was referring to included, among others, the aborigines and the blacks. What the racist Adolf Hitler visited upon the Jews is common knowledge to everyone except for a number of half-witted, Jew-hating Muslims (like Farrakhan, Ilhan Omar and buffalo ass face Jayapal) and other holocaust deniers. But what is not common knowledge, nor is even whispered in our educational system or by our lying, Democrat media (but I repeat myself), is that it was the theory of evolution that gave him the "scientific" basis, as well as the "moral" imperative, to carry out his horrific crime against humanity.

And then there is the founder of Planned Parenthood, Margaret Sanger, another devout Darwinist who we will deal with in more depth

---

254 (Quote from *The Mysterious Islands: A Surprising Journey to Darwin's Eden,* a film by the Erwin Brothers)

255 (From *Ontology and Phylogeny,* 1977, by leading evolutionist Stephen Jay Gould. As quoted in *The Mysterious Islands: A Surprising Journey to Darwin's Eden,* an Erwin Brothers film)

in the following chapter (see book "All the Secrets"). Like Hitler, she was committed to the idea of improving the human race by helping evolution along, in her case by eliminating the Negro. In a December 19th, 1939 letter to Dr. Clarence Gamble she said this, "We don't want the word to get out that we want to exterminate the Negro population."[256] And keep in mind that she definitely *wanted* to exterminate the Negro population; she just didn't want the word about it to get out. And where are most Planned Parenthood clinics located in this country today? You guessed it, poor black neighborhoods. 40% of all babies mutilated to death every year are black babies, yet blacks are only 12% of the population. Can you say, "slow genocide?" Hey Trump hater, LeBrainless James, need a calculator? And Hillary Clinton said, "Margaret Sanger is my hero!!" And who do 90% plus of blacks go out and vote for religiously every two years? You guessed it, Planned Parenthood and the Democrat Party, and therefore the slow purposeful genocide of their own race. Ouch! Wake up, my brothers and sisters! You've been lied to all of your life by a filthy Demonrat media, educational system, entertainment industry and sadly the vast majority of black churches who Sanger bought off in the 1960's. Time to start listening to Candace Owens…

PETA

Another of the unintended consequences of the acceptance of Darwinism throughout our society—specifically the belief that man is a mere animal, no different than any other animal or creature along the "evolutionary" ladder—is that it opened the sanatorium doors and let out a uniquely-modern lunatic fringe. Evolution serves…

> …"as the philosophical foundation for the animal rights movement. After all, if man is simply another evolving animal, what gives him rights superior to bugs, and birds, and bats? This is the point frequently made by the leaders of People for the Ethical Treatment of Animals [PETA], perhaps the most visible and outspoken element of the animal rights community."[257]

---

256 (Quoted from *The Mysterious Islands: A Surprising Journey to Darwin's Eden,* an Erwin Brothers Film)

257 (From *The Mysterious Islands* video featurette, "Galapagos Whaling Controversy: A Christian Perspective," by Doug Phillips)

PETA president, Ingrid Newkirk, actually thinks that the holocaust should be equated with the horror of eating fried chicken…

> "Six million Jews died in concentration camps, but six billion broiler chickens will die this year in slaughterhouses."[258]

(Lucky for Frank Purdue the Nuremburg trials are over.)

And that humans and rats are one and the same…

> "There is no rational basis for saying that a human being has special rights. A rat is a pig is a dog is a boy. They're all mammals."[259]

Ms. Newkirk is a rat is a pig is a dog is a flipping idiot. But that is the result of being brainwashed in the evolutionary lies of the wicked one. God help her.

And then there is this gem, from another animal rights lunatic…er… advocate…

> "The life of an ant and the life of my child should be granted equal consideration."[260]

And with that said, social services should remove his child immediately. (Give him an ant farm.)

Conversely, this is what God says, the actual Creator of life…

"And God said, Let us make man in our image, according to our likeness: and let them <u>have dominion over</u> the fish of the sea, and over the birds of the air, and over the cattle, and over all the earth, and over every creature that moves upon the earth. So God created man in his own image, in the image of God he created him; male and female he created them. And God blessed them, and God said unto them, Be fruitful, and multiply, and replenish the earth, and subdue it: and <u>have dominion over</u> the fish of the sea, and over

---

258   (Ingrid Newkirk, President and Founder, PETA)

259   (Ingrid Newkirk, President and Founder, PETA, People for the Early Treatment of Asininity)

260   (Michael Fox, Vice President, The Humane Society- USA. And that mentally disturbed moron should be granted consideration for a strait jacket, and the immediate removal of his children.)

the birds of the air, and over every living thing that moves upon the earth." (Genesis 1:26-28)

"What is man that you are mindful of him, and the son of man, that you care for him? For you have made him a little lower than the angels, and have crowned him with glory and honor. You have made him to have dominion over the works of your hands; you have put all things under his feet: all sheep and oxen, and the beasts of the field, the birds of the air, and the fish of the sea, and whatsoever passes through the paths of the seas." (Psalms 8:4-8)

> "From these two passages we learn two important points: first, that man alone was made in the image of God. And second, that man was given dominion over all the creatures. ...[Having dominion] doesn't mean that man can abuse or exploit creation as he pleases. Nor should we demonstrate cruelty to animals. ... 'The righteous man cares for the life of the beast.' (Proverbs 12:10) ...We ought to be wise stewards of the rich resources we have been given, and with thankful hearts use them for the good of mankind.
>
> "We should also recognize that the notion that it's criminal and cruel to harvest animals because they are big, or cute, or fluffy, or beautiful is a neo-Darwinian philosophy born out of self-loathing for humanity. ...It's not accidental that many of the most vocal advocates for animal rights are also defenders of the rights of women and physicians to kill human babies within the womb... Save the whales. Kill the babies. It makes perfect sense in a Darwinian world. And it's this type of perverse perspective that inspires leaders of this movement to compare the genocide of Jews by Adolph Hitler to the so-called sin of eating fried chicken."[261]

PETA and our animal "rights" friends have passionately and self-righteously immersed themselves and their lives in a crusade that is all built upon a lie; the idiotic fantasy of evolution. What a waste of a lifetime. They need to immerse themselves in the truth, and the real righteousness that can only come through faith in Jesus Christ, life's real Creator. Pray for them. And pray that the bald-faced, scientifically-impossible lie that enslaves their

---

261 (From *The Mysterious Islands* video featurette, "Galapagos Whaling Controversy: A Christian Perspective," by Doug Phillips)

minds and their hearts will soon be exposed nationally and internationally for what it is.

"Where were you, stupid, when I laid the foundations of the earth? Tell me, if you have understanding. …On what are its foundations fastened? Or who laid its cornerstone; when the morning stars sang together, and all the sons of God shouted for joy? …When I made the clouds its garment, and thick darkness its swaddling band" (Job 38:4, 6-7, 9)

(I know, "stupid" isn't really in there. I just threw it in to make sure you were paying attention.)

"Theistic" Evolution

Evolution's bastard child. We need to mention it here, briefly, because sadly many millions of uninformed Christians (usually of the faux variety) have been bamboozled into accepting this farcical amalgamation that attempts to blend belief in God and His Word with the scientific "fact" (read: lie) of evolution. It's a fool's bargain, and one that is completely unnecessary for any believer because the only thing the lie of evolution needs to be amalgamated with is the manure pile. Evolution poses no challenge to belief in God, but since so many uninformed Christians remain ignorant of this fact, they have "comforted" themselves by believing in its theistic version. *Theistic evolution* goes something like this, "Well, we *know* that evolution is *science* and therefore it is a *fact*, but we also believe in God, so we believe God <u>helped</u> evolution along the way. It was God who was behind all of the evolving over all of the hundreds of millions of years."

As such, theistic evolution is possibly more stupid than the actual theory itself. It is a pathetic attempt to reconcile the God of Creation with the lie of evolution. It is insulting, and it only exists because so many Christians, and other believers in God, have never been educated as to the scientific impossibility of the bald-faced lie that macroevolution is. God is not an idiot. (Man is another story entirely.) God is all-powerful, and all-intelligent. The idea that he had to muddle around for millions of years trying to change one completely viable and incredibly engineered organism into another, rather than just making them all at once, is patently absurd. I mean, as so-called Christians, who are we going to believe? Brainwashed Darwinists, faux scientists all, who weren't even there, who haven't the slightest idea what they are talking about, or God who actually was there at Creation? (Take your time.) Indeed, theistic evolution was merely created so those Christians who had bought into the scientifically absurd fiction that is evolution, could try to square it with the Word of God. It doesn't work. Either Satan is right (which of course he isn't) or the Bible is absolutely true, and exactly what it says it is over three thousand times within its pages and that would be, <u>written by God</u>! (For further, undeniable proof that the Bible was indeed written by God see chapter 6 "The Word of God" in the larger work "All the Secrets of the Universe: All the Ones worth Knowing. A Graduate Course on the truth.") And God is not a liar. Evolutionists, however, are. And so too are "theistic" evolutionists. And both are also calling God a liar. Wake up! Learn the truth. If you're going to call yourself a Christian, stop aiding and abetting the enemy. (Something that many on the far-left who call themselves Americans might also learn.)

Either Genesis is scientifically true or it is not. If it is not then the Bible is a lie and Jesus quoted from a lie... and his death on the cross to pay the price for our sins, his resurrection from the dead three days later, and the salvation he offers are all highly suspect. And that is just the way the evil one wants it. The theory of evolution is Satan's genesis. It is his origins account. It is his lie. And he is quite content in having foolish, uneducated "Christians" believing in his lie instead of God's accurate account, and to then try and make his lie "compatible" with the Bible by coming up with the utter stupidity known as theistic evolution. God have mercy on us idiots.

PART IV

IN THE BEGINNING

"These are the generations of the heavens and of the earth when they were created, in the day that the Lord God made the earth and the heavens." (Genesis 2:4)

The first chapter of the first book of the Bible records the sequence of events that culminated in God's finished work of creation: this universe of stars and galaxies, this solar system with our water-planet earth, and all of life upon it. But not only is it obvious that a great amount of specifics and details have been left out, have been left unsaid in this first chapter of Genesis, but there has also been a great deal of confusion over the meaning of what actually has been said. For example, what exactly do phrases like "the earth was without form and void," and "darkness was upon the face of the deep," and "the Spirit of God was hovering over the face of the waters," actually mean? What was "the deep," and what was "the face of the waters" that "the Spirit of God was hovering over?" Is there any possible way for us to accurately answer these questions, and to come to a clear understanding of what literally happened in those first few days?

I think there is, and that the answer may have come by way of a book by D. Russell Humphreys, a Ph.D. in physics who for years "worked for Sandia National Laboratories (New Mexico) in nuclear physics, geophysics, pulsed-power research, and theoretical atomic and nuclear physics."[262] It is titled, *Starlight and Time: Solving the Puzzle of Distant Starlight in a Young Universe,* and his unveiling of the mysteries of Genesis, using his extensive knowledge of physics and astronomy, can be found in chapter 2, "Creation Week: A Possible Scenario."

Dr. Humphreys begins his analysis on Day One (Genesis 1:1-5):

---

262 Humphreys, *Starlight and Time,* Back cover.

156

"In the beginning God created the heaven and the earth. And the earth was without form, and void; and darkness was upon the face of the deep. And the Spirit of God was hovering over the face of the waters." (Genesis 1:1-2)

According to Dr. Humphreys calculations, God began by creating both 3 dimensional space ("the heavens") a billion light years in diameter, and a sphere of liquid water which contained all of the matter of the universe ("the earth") over 2 light-years in diameter at the center of this "large 3-D space." It is fitting that the Creator should begin this material world with water, as water is life. "And he said unto me, It is done. I am Alpha and Omega, the Beginning and the End. I will give unto him who thirsts of the fountain of the water of life freely." (Revelation 21:6)

> "The ball [of matter-water] is greater than two light-years in diameter, large enough to contain all the mass of the universe… Two light-years is surprisingly small compared to the later size of the universe, but it is still huge (about 12 trillion miles or 20 trillion kilometers) compared to us, being more than a thousand times greater than the diameter of our solar system. Imagine floating on the face of the deep and gazing down into its unimaginable depths! …The earth at this point is merely a formless, undefined region of water at the center of the deep, empty of inhabitant or feature. The deep is rotating slowly and there is no visible light at its surface."[263]

What other explanation fits with this description in Genesis of the future, physical earth being "without form, and void," which would have been at that point just an infinitesimal spec in the center of this immense ball of water-matter? And what other explanation fits with the phrase "the deep" other than a massive sphere of water? "The Spirit of God was hovering over the face of the waters" because that was all there was, water. And then there is this in the New Testament, "By the word of God the heavens existed long ago and **the earth was formed out of water and by water**." (2 Peter 3:5) That kind of nails it right there. Dr. Humphreys continues:

> "In the early 1980's, I based a theory about the origin of planetary magnetic fields on the possibility that the earth and other bodies in the solar system were originally created as pure water. The theory has been remarkably

---

263 Humphreys, *Starlight and Time,* 32-33.

successful, even to the point of correctly predicting the Voyager spaceprobe's measurements of the magnetic fields of the planets Uranus and Neptune.

"The theory could not work with the present elements composing the solar system bodies, but only with water as the original material. Thus it seems that transformation of water on day one (the modern word is "nucleosynthesis") is a distinct biblical and scientific possibility.

…"If the "nucleosynthesis" scenario above is correct, then it means that at the instant of creation, the earth was merely a small region of water at the center of a much larger ball of water, the deep. That region had no distinguishing marks and was empty of any other kind of matter. This, I suggest, is the meaning of the much-discussed phrase, "formless and void" [in Genesis 1:1-2]."[264]

"Because the enormous mass of the whole universe is contained in a ball of (relatively) small size, the gravitational force on the deep is very strong, more than a million trillion "g"s. This force compresses the deep very rapidly toward the center, making it extremely hot and dense. The heat rips apart the water molecules, atoms, even the nuclei into elementary particles."[265]

"And God said, Let there be light: and there was light." (Genesis 1:3)

"Thermonuclear fusion reactions begin, forming heavier nuclei from lighter ones and liberating huge amounts of energy. As a consequence, an intense light illuminates the interior, breaking through to the surface and ending the darkness there."[266]

"And God saw the light, that it was good: and God divided the light from the darkness." (Genesis 1:4)

"As the compression continues, gravity becomes so strong that light can no longer reach the surface, re-

---

264 Humphreys, *Starlight and Time*, 73.

265 Humphreys, *Starlight and Time*, 33.

266 Humphreys, *Starlight and Time*, 33.

darkening it. Psalm 104:2, "Covering Thyself with light as with a cloak," in context appears to refer to Day One. This suggests to me that at this point the Spirit of God, "moving [or hovering] over the surface of the waters" (Genesis 1:2), becomes a light source, in the same way as He will again become a light source at a future time (Revelation 21:23, 22:5). This would give the deep a bright side and a dark side, thus dividing light from darkness and inscribing "a circle on the face of the waters, at the boundary of light and darkness." (Job 26:10)."[267]

"And God called the light Day, and the darkness he called Night. And the evening and the morning were the first day." (Genesis 1:5)

"The deep speeds up its rotation as the compression continues, as a whirling ice skater speeds up when she brings her arms inward. We can imagine a reference point on the surface rotating around to the dark side and continuing further around to the bright side again, marking off evening and morning. Rough calculations show that all of the events from the beginning to this point had to take place in a very short time... To calculate the time exactly would go beyond the frontiers of modern relativity, but I suspect that... [it was] about 24 hours from the instant of Creation to the end of Day One."[268]

I would have to agree, mainly because it is abundantly clear from linguistic studies of the Bible that during Creation week God was talking about *literal 24-hour days*, and not some vast eons of time as proposed by theistic evo-delusionists.

Dr. Humphreys continues his analyses of Genesis on Day Two (Genesis 1:6-8):

"And God said, Let there be an expanse [firmament] in the midst of the waters, and let it divide the waters from the waters. And God made the expanse, and divided the waters which were below the expanse from the waters which were above the expanse: and it was so." (Genesis 1:6-7)

---

267   Humphreys, *Starlight and Time,* 33-34.

268   Humphreys, Starlight and Time, 34.

"By direct intervention... God began stretching out space, causing the ball of matter to expand rapidly. [Who covers *thyself* with light as *with* a garment: **Who stretches out the heavens like a curtain**" (Psalms 104:2)] ...He marks off a large volume, the "expanse" ("firmament" in the KJV) within the deep, wherein material is allowed to pull apart into fragments and clusters as it expands.

..."Normal physical processes cause cooling to proceed as rapidly as the expansion. Heat waves are stretched out to much longer wavelength as a relativistic consequence of the stretching of space. Eventually the stretched-out waves will become the cosmic microwave background radiation.

"Matter beneath the expanse expands until the surface reaches ordinary or present temperatures, becoming liquid water underneath an atmosphere. God collects various heavier atoms beneath the surface (formed from fusion reactions as mentioned earlier) and constructs minerals of them, laying "the foundations of the earth" (Job 38:4), i.e., its core and mantle. Gravity at the surface drops to normal or present values. Out in the expanse, matter is drawn apart, leaving irregular clusters of hydrogen, helium, and other atoms formed by the nuclear processes of the first day."[269]

"And God called the expanse Heaven. And the evening and the morning were the second day." (Genesis 1:8)

..."These heavens are interstellar space. Since the sun has not yet been created, the Spirit of God continues to be the light source close to the rotating waters below, giving them a light and dark side. The expansion started at the beginning of this day will continue until at least the end of the fourth day."[270]

Day Three...

"And God said, Let the waters under the heaven be gathered together into one place, and let the dry land appear: and it was so." (Genesis 1:9)

---

269   Humphreys, *Starlight and Time*, 34-36.

270   Humphreys, *Starlight and Time*, 36.

"Rapid radioactive decay occurs... possibly as a consequence of the rapid stretching out of space. The resulting heating forms the earth's crust and makes it buoyant relative to the mantle rock below it, causing the crust to rise above the waters, thus gathering the waters into ocean basins. I hypothesize that rapid volume cooling of molten rock deep within the earth also occurs, again as a result of their rapid expansion of space... solidifying the rock. ...There are no stars yet, only clusters of hydrogen, helium, and other atoms left behind in the expanse by the rapid expansion."[271]

Day Four...

"And God said, Let there be lights in the firmament of the heaven to divide the day from the night; and let them be for signs, and for seasons, and for days, and years: And let them be for lights in the firmament of the heaven to give light upon the earth: and it was so. And God made two great lights; the greater light to rule the day, and the lesser light to rule the night: he made the stars also. And God set them in the firmament of the heaven to give light upon the earth, And to rule over the day and over the night, and to divide the light from the darkness: and God saw that it was good. And the evening and the morning were the fourth day." (Genesis 1:14-19)

"During this ordinary day [the fourth day] as measured on earth, billions of years' worth of physical processes take place in the distant cosmos. [This is due both to the expansion of space and matter, and gravitational time dilation.] In particular, gravity has time to make distinct clusters of hydrogen and helium atoms more compact. Early on the fourth morning, God coalesces the clusters of atoms into stars and thermonuclear fusion ignites in them. The newly-formed stars find themselves grouped together in galaxies and clusters of galaxies. As the fourth day proceeds on earth, the more distant stars age billions of years, while their light also has the same billions of years to travel to the earth. While the light is on its way, space continues to expand, relativistically stretching out the light waves ...and shifting the wavelengths toward the red side of the spectrum. Stars which are now farthest away

---

271 Humphreys, *Starlight and Time,* 36-37.

have the greatest red shift, because the waves have been stretched the most. This progressive redshift is exactly what is observed."[272]

We will skip to Day Six…

"And God saw every thing that he had made, and behold, it was very good. And there was evening and there was morning, the sixth day." (Genesis 1:31)

> "God stops the expansion… before the evening of the sixth day. Therefore, Adam and Eve, gazing up for the first time into the new night sky, can now see the Milky Way, the Andromeda galaxy, and all the other splendors in the heavens that declare the glory of God."[273]

Wow! "Lord of all creation, of water earth and sky. The heavens are your Tabernacle. Glory to the Lord on high. God of wonders beyond our galaxy. You are holy, holy. The universe declares your Majesty. You are holy, holy. Lord of Heaven and Earth."[274]

He truly is a "God of Wonders!" The "Lord of all creation, of water earth and sky." How awe inspiring it is to catch a glimpse of how it *really* all began! Not from some mindless, stupid, accidental explosion, ill-conceived by those who ignorantly reject God and any consideration of His creative work. Who "professing themselves to be wise, became fools. …Who changed the truth of God into a lie, and worshipped and served the creature more than the Creator." (Romans 1:22, 25) An explosion that couldn't create a damn thing, much less this universe of incredible beauty, exact physical laws and precisely-tuned forces, all spoken into existence by the holy "Lord of Heaven and Earth" who possesses an intelligence that stands in infinite contrast to those *faux scientists* and *intellectual elites* who refuse to acknowledge him as their Creator. And what could be more perfect then for He who *is* the Water of Life to have begun his universe by creating water? "My people have committed two evils; they have forsaken me the fountain of living waters, and hewn them out cisterns, broken cisterns that can hold no water." (Jeremiah 2:13) We as

---

272 Humphreys, *Starlight and Time*, 37-38.

273 Humphreys, *Starlight and Time*, 38.

274 (From song *God of Wonders,* by Third Day. …Check out the video on YouTube)

a nation and a people have forsaken our own Creator, and worship random chance, a "broken cistern," in his stead.

And then he stretched the whole thing out, and changed the water into all of the known and unknown elements, and made stars, countless numbers of them and formed them into billions of galaxies. The universe truly declares His majesty! This "God of wonders beyond our galaxy," beyond all galaxies, beyond all time and space. And at the center of it all He made this tiny solar system with this little water-planet-earth and orbiting-moon system uniquely designed to sustain and nurture life in the midst of the cold dark expanse of outer space; a single planet made up of an infinitesimal amount of the universes matter, yet of more significance than all the rest of the universe combined; where he would breathe all the living into being and crown this planet teeming with all variety and splendor of life with man, made in His image and His likeness. "Early in the morning, I will celebrate the Light,"[275] because he made us from the light of His own Conscious Spirit, to serve Him and glorify Him, and to have a part in letting "the earth be filled with the knowledge of the Glory of the Lord, as the waters cover the sea." And behold He saw that it *was* very good.

"Thus says the Lord, who stretches out the heavens, lays the foundation of the earth, and forms the spirit of man within him…" (Zechariah 12:1)

How certain can we be, though, that Dr. Humphreys unveiling of Creation Week is completely accurate? Roughly 85.67 %. (And I did that without a calculator.) Seriously though, this is just one scientist, a brilliant one to be sure, and one unfettered by the shackles of an absurd, idiotic theory that leads its followers on a wild goose chase looking for origins in all the wrong places. And this is that one scientist's interpretation of Genesis. It is his version of the specifics of creation week not spelled out in God's Word, using the known laws of the universe and those parts of the general theory of relativity that appear to be accurate. It is not "written in stone." But God's version *is* written in Stone. It is written in stone by the Chief Cornerstone, and it accurately describes what happened in the beginning. But at the same time much is obviously left unsaid, and because of that much has also been left wrapped up in mystery for thousands of years. A mystery that has defied an accurate and scientifically sound interpretation, perhaps until now. Whether Dr. Humphreys has absolutely accurately unveiled that mystery, I can't say. You decide. (I would think, though, that even at the worst, he has

---

275   (From song *God of Wonders*, by Third Day)

come very close.) But until some non-deluded, non-faux scientist comes up with a better and more accurate scientific interpretation, it works for me. Dr. Hawking, are you up to the challenge? Can you leave the scientific absurdity of big bang cosmology alone long enough to search out the truth? The world of true science awaits your decision; with bated breath. (Obviously this was written before he passed. RIP)

"And, you, O Lord, in the beginning have laid the foundations of the earth; and the heavens are the work of your hands." (Hebrews 1:10)

CONCLUSION

"God, who at various times and in sundry manners spoke in time past unto the fathers by the prophets, has in these last days spoken unto us by his Son, whom he has appointed heir of all things, by whom also he made the worlds. Who being the brightness of his glory, and the express image of his person, and upholding all things by the word of his power, when he had by himself purged our sins, sat down on the right hand of the Majesty on high." (Hebrews 1:1-3)

Science is based upon observation. No one has ever observed macro evolution, nor will they. But there is someone who observed Creation. He was the world's first Scientist, and science can know that His witness is true. It would be wise for America to wake up to this reality.

The theory of evolution's preeminence in our educational system has everything to do with the Democrat Left's fascist control over that system as we have already stated, but the *roots* of evolution's total preeminence can be traced back to the height of the cold war, and to the earliest years of the "space race:"

> "The USSR's ability to place a satellite in orbit raised a cry of alarm about the "Space Gap" between superpowers—how had we let American resolve become so flabby? One result was a federal imperative to improve high-school science education—which led, not incidentally, to evolution instruction being mandated across the board."[276]

Well, we made it to the moon, with no thanks to the "across the board mandating" of the teaching of a scientifically-impossible lie, and the Russians still haven't. But before we start overly congratulating ourselves, the fact remains that we are too scientifically ignorant and uneducated to mandate teaching our school children the truth about how the moon *got* there in the first place. The *space race* is over, but the question that still remains is, *Why*

---

276   (From *One Large Misstep,* by Janie B. Cheaney, article in World Magazine, Aug. 2009)

are we so stupid as to continue to allow bald-faced, scientifically-impossible lies to be taught, under the guise of science, and as if they were fact, to our naïve and unsuspecting school children in our colleges, universities, elementary and high schools?

And the answer of course is that the lies of Satan cover this earth like a blanket and people believe whatever they are taught. It is so easy for us here in America to look at all of those poor, unfortunate Muslim jihadists who actually thought they were going to go to an eternal paradise replete with 72 virgins for their unending pleasure and use, by flying a plane into a tall building and killing thousands of innocent men, women and children. And we cry out, "HOW COULD ANYBODY BE THAT STUPID?" Yet we should be shouting that into a mirror, because the main reason that many of us <u>don't</u> believe that uniquely Islamic stupidity is because <u>we weren't taught it.</u> But the bald-faced, equally-as-stupid, evolutionary lies we <u>have</u> been taught, those we <u>do</u> believe. (And for the record, I am not pointing the finger because I believed them too, before I was brought from the lies of Satan's genesis to the knowledge of the truth. ..."He rescued me from the power of darkness, and transferred me into the kingdom of his dear Son." Colossians 1:13-14)

When we look at the stunning miracle that is life—the computing marvel that is the human brain with its hundred trillion electrical connections; the sixty thousand miles of blood vessels that course through our bodies; the astonishing imagination that designed the metamorphosis of insects and the resulting breathtaking beauty of many of its end results such as the butterfly; the creativity and unfathomable intelligence that could place an entire library of coded information in a seven-foot-long strand of DNA and somehow coil it up and put it into a living cell a billionth of a cubic inch in size; the mind-boggling, nanobot, engineering intricacy of the living cell that functions and reproduces for years without any conscious intervention—it is all so absolutely overwhelming. It is almost beyond belief that it all even exists. Yet He merely spoke, and it all came into being. By the breath of his mouth. "By the word of the Lord were the heavens made, and all the host of them by the breath of his mouth" (Psalm 33: 6) And yet the equally amazing and astronomically sad thing is that instead of glorifying and honoring and praising and thanking the God of Wonders, the Lord of all Creation, for bestowing this incomprehensible miracle gift upon us, we spit in his face by giving all the credit to an idiot, random accidental chance, and to a bald-faced liar, the devil. Again, may God have mercy on all of us morons.

"You are worthy, O Lord, to receive glory and honor and power; for you have created all things, and for your pleasure they exist and were created." (Revelation 4:11)

Let us give glory to the God of Creation, the author of the Bible, who inspired Moses to record in the Book of Genesis the scientifically accurate history of the creation of the world. Who, unlike our friend Beelzebub who has inspired a competing genesis account, is <u>not</u> a liar. Who spoke all the universe into existence during a six-day period six thousand years ago; and who breathed all the living into being during the same week; and who then "rested the seventh day, wherefore that same Creator Lord blessed the Sabbath day, and hallowed it." (Exodus 20:11) And <u>that</u>, unlike evolution, <u>is</u> a scientific fact!

"Lift up your eyes on high, and behold who has created these things" (Isaiah 40:26)

"He was in the world, and the world was made by him, and the world knew him not." (John 1:10) Because the world did not care to know Him, nor to be bothered with His truth. For if it did care to know him and the truth, both would be very easy for them to find. As we said in "All the Secrets," the universe is not the one keeping certain very important things secret from us. We are. We just refuse to hear the truth, so certain things remain secret, like the scientific fact that evolution really is a joke.

"So I say unto you, ask, and it shall be given you; seek, and you shall find; knock, and it shall be opened unto you." (Luke 11:9)

"And you shall seek me, and find me, when you shall search for me with all your heart."(Jeremiah 29:13)

"And you shall love the Lord your God with all your heart, and with all your soul, and with all your mind, and with all your strength." (Mark 12:30) And the "Lord your God" and the Truth are one and the same… "I am the truth." (John 14:6)

"Fear God and give glory to him, for the hour of his judgment has come; and worship him who made heaven and earth, the sea and the fountains of water." (Revelation 14:7)

"But sanctify the Lord God in your hearts: and be ready always to give an answer to every man who asks you a reason for the hope that is in you with meekness and fear." (1 Peter 3:15)

Now we know the truth about the evolution-creation debate. There is none. All that remains is a battle between the idiotic, asinine, insanely stupid lies of the great deceiver and the truth of God our Creator. And in many ways that truth is extremely privileged information, and you and I are extraordinarily blessed that it has been revealed to us. It is another incredible "secret of the universe;" and *secret* only because it has been systematically and deliberately kept hidden from the majority of people today by the great deceivers absolute control over the means of disseminating information in our culture. There are countless multitudes of people who have not received this blessing. And you and I both will be required to answer for what has been freely given us. ("For unto whomsoever much is given, much shall be required." Luke 12:48) Will we hide our light under a bushel, or will we shine it on a hill? (See the parable of the talents in Matthew 25:14-30) The light of the truth of the glorious gospel of Jesus Christ has been hid from this world in many ways because of the darkness bestowed upon it by Satan's genesis. It is therefore incumbent upon anyone who has been blessed with the knowledge of this truth to pass it along to as many as possible. To this end, whether you are young or old, teacher or student, in a position of great power and influence or of humble means, may God give you the courage and the boldness to do so.

"I can do all things through Christ who strengthens me." (Philippians 4:13)

"For by him were all things created that are in heaven and that are on earth, visible and invisible, whether thrones or dominions or principalities or powers. All things were created by him and for him." (Colossians 1:16)

Amen.

Bibliography and Recommended Reading:

- John D. Morris, Ph.D., *The Young Earth: The Real History of the Earth — Past, Present, and Future* (Green Forest, AR: Master Books, 2007)

- John D. Morris, Ph.D., *Dinosaurs, the Lost World, and You*

- Scott M. Huse, *The Collapse of Evolution* (Baker, 1986)

-*In The Beginning: Compelling Evidence for Creation and the Flood,* by Walt Brown, Ph.D.

- Walt Brown, Ph.D., *In The Beginning: Compelling Evidence for Creation and the Flood* (Phoenix: Center for Scientific Creation, 2008)

- Tom DeRosa, *Evidence for Creation: Intelligent Answers for Open Minds* (Powder Springs, GA: Creation Book Publishers, 2001)

- *The Mysterious Islands: A Surprising Journey to Darwin's Eden*, An Erwin Brothers Film

- Kyle Butt and Eric Lyons, *Dinosaurs Unleashed: The True Story about Dinosaurs and Humans* (Montgomery, AL: Apologetics Press, 2004)

- Dr. Michael J. Behe, *The Edge of Evolution: The Search for the Limits of Darwinism* (New York: Free Press, 2007)

- David Berlinski, *The Devil's Delusion: Atheism and its Scientific Pretensions,* (New York: Crown Publishing Group, 2008),

- Michael Denton, *Evolution: A Theory in Crisis,* (Chevy Chase, MD: Adler and Adler, 1986)

- Michael J. Behe, *Darwin's Black Box: The Biochemical Challenge to Evolution,* (New York: Free Press, 2006)

- *Expelled: No Intelligence Allowed,* a documentary movie narrated by Ben Stein

- D. Russell Humphreys, Ph.D., *Starlight and Time: Solving the Puzzle of Distant Starlight in a Young Universe* (Green Forest, AR: Master Books, 2006),

- Jonathan Sarfti, *The Greatest Hoax on Earth? Refuting Dawkins on Evolution* (Powder Springs, GA: Creation Book Publishers, 2010)

- Dr. Don DeYoung, *Thousands… Not Billions: Challenging an Icon of Evolution, Questioning the Age of the Earth* (Green Forest, AR: Master Books, 2006)

- Ken Ham, *The LIE: Evolution* (Green Forest, AR: Master Books, 2007)

- Tom DeRosa, *Evolution's Fatal Fruit: How Darwin's Tree of Life Brought Death to Millions* (Fort Lauderdale: Coral Ridge Ministries, 2006)

- Dr. Jason Lisle, *Taking Back Astronomy: The Heavens Declare Creation* (Green Forest, AR: Master Books, 2006)

- Tom Vail (Canyon river guide, who wrote and compiled...), *Grand Canyon: a different view* (Green Forest, AR: Master Books, 2003)

- James I. Nienhuis, *Ice Age Civilizations* (Houston: Genesis Veracity, 2006)

- Ron J. Bigalke Jr. Compilation Editor, *The Genesis Factor: Myths and Realities* (Green Forest, AR: Master Books, 2008)

- Marvin L. Lubenow, *Bones of Contention: A Creationist Assessment of Human Fossils* (Grand Rapids: Baker Books, 2007) (**NOT** the one by Roger Lewin!)

- Phillip E. Johnson, *Darwin on trial,* (Washington DC: Inter Varsity Press, 1993) (Also see: *Darwin Found Guilty, Darwin Sentenced, Darwin Executed* and, the final one in the series, *Darwin Buried in Bill Maher's Back Yard,* by I. M. Sighents, Ph.D.)

- Guillermo Gonzalez and Jay W. Richards, *The Privileged Planet: How Our Place in the Cosmos is Designed for Discovery* (Washington DC: Regnery Publishing, 2004)

- Dr. Werner Gitt, *In the Beginning was Information: A Scientist Explains the Incredible Design in Nature* (Green Forest, AR: Master Books, 2005)

- Fazale Rana, Ph.D., *The Cells Design: How Chemistry Reveals the Creator's Artistry* (Grand Rapids: Baker Books, 2008)

- Is Genesis History? (Answer: YES it is.) A Documentary by Thomas Purifoy Jr.

ADDENDUM 1

Table of Contents, Introduction and Conclusion to "All the Secrets of the Universe (All the Ones worth Knowing): A Graduate Course on the Truth."

****

All the Secrets Of the Universe

(All the ones worth knowing)

A Graduate Course on the Truth

By

Danny Erhardt

Copyright 2009

Book II   FIRE

<u>Book III   WATER</u>

Book IV   EARTH

INTRODUCTION

"Evil men do not understand judgment:
but they that seek the Lord understand all things."
(Proverbs 28:5)

The author of those words, written thousands of years ago, would be ridiculed if he spoke them today on any one of our college or university campuses. For the ideological-left elites who control the intellectual discourse there hold the idea of "the Lord" in complete contempt, much less the idea that one could come to an "understanding of all things" merely by "seeking Him." But their ignorant scorn notwithstanding, the question remains, was that ancient author right? Did he know what he was talking about? Is it possible to rise above all of the confusion, chaos and senselessness that covers this planet and come to a clear understanding of what is going on, and why, just by seeking the Lord? (And it should be noted that *the Lord* in this context is understood to be the all-powerful, all-knowing and infinitely intelligent Creator of this universe.[277] It is also understood that the Lord and the truth are one and the same, that they are inseparable; for the Lord <u>is</u> the truth. So in seeking Him one is also seeking the truth.)

The truth can be known

So, with all that said, *is* it possible to figure everything out about this life: "this sore travail that God has given to the sons of man to be exercised herewith?" (Ecclesiastes 1:13) Can we know for a certainty the answer to all of life's fundamental questions; those that have puzzled mankind, inspired poets, and tied the minds of philosophers and academics in knots for millennia? Or is anybody's guess just as good as anyone else's? Is anyone's opinion as valid

---

277   (And not to jump the gun but as we will see in chapters 6 and 10, and despite what we were all taught, that unlike the scientifically-impossible and imaginary creators of random accidental chance (the theory of evolution) and a real big explosion (big bang theory) true science itself confirms that the Lord is the actual Creator of this Universe as well. Go figure.)

as anyone else's? Can we know what really caused the universe to come into being; why it does exist; how life really originated; why we are here and what is our purpose in being here (*what's the point?*); which ideologies are true and which are false; which religious claims are true and which are false; "I SAY HEY, HEY, HEY, HEY, WHAT'S GOIN' ON"[278] right now on this panicked planet, and why; and where are we going, if anywhere, after we depart this life? Is it possible to know the actual answers to these and other related questions? Not just to hold an opinion on each one, but to actually uncover their true answers, and thereby come to a clear *understanding of all things?*

Well as strange as it may seem, and despite what the "current wisdom" of this day assumes, the answer is yes. And even though the reader might be skeptical about this possibility at this point, we can assure you that if you follow along through the pages of this book, and if you determine to seek the truth for yourself, with all of your mind and all of your heart and all of your soul, then you will find by the end, and hopefully well before it, that you have come to agree with this conclusion. You will find what millions who have traveled this path before you have already found. That it is possible to leave behind the confusion, fear and anxiety that come from not knowing the truth; from having been taught any number of false assumptions, false beliefs, false religions, false ideologies and unscientific theories. And that by seeking the truth and leaving the lies far behind, it is possible to make sense of all this madness. That it is possible to know what is going on and why. That it *is* possible to understand all things.

"Come unto the truth, all you who labor and are heavy laden, and He will give you rest. Take His yoke upon you, and learn from Him; for He is gentle and humble in heart: and you shall find rest for your souls. For His yoke is easy, and His burden is light." (Matthew 11:28-30, in the third person.)

One of the side benefits of seeking the truth, besides that which comes from any increase in knowledge (it being power if applied correctly), is peace to the mind and rest to the soul. And in this world of unending war, political turmoil, economic collapse, catastrophic tsunamis, horrific tornadoes, violent Islamic insanity, and socialist takeovers here at home by reprobate minded, literally insane, America despising, racist, black-baby exterminating, false-accusing Democrats, no price can be put on that. ☺

The more I know, the less I understand

----

278   (As *screamed from the top of her lungs* in the song "What's up?" by 4 Non Blondes)

There is a difference between knowledge and understanding. (One is knowledge. The other is understanding.) (Shazaam.) There is also an infinite difference between knowing all things and understanding all things. The former is impossible. You would have to have the mind of God. Imagine that you possessed all of the information from every book in every home, school and library across the face of this earth, as well as all of the data stored in the memory and hard drives of every computer on the planet, in addition to every memory and every piece of information in the mind of every human being alive today and everyone who has ever lived, and then multiply that by ten billion to the ten billionth power; and even then you would still only have a mere infinitesimal fraction of the totality of all knowledge.

Understanding all things, fortunately, is a tad easier. For only a relatively limited and also easily accessible volume of knowledge is required to comprehend clearly what's going on in this world, and why it is, and to discover where the truth lies and where it does not. This journey to understanding requires the ability to separate fact from fiction, truth from lies, and that is an innate ability which is also readily available to any and all who will choose to use it. So don't be fooled by the "conventional wisdom" of this world that the answers to all of life's fundamental moral, philosophical and theological questions are not attainable in the absolute; that this is a realm left to the world of opinion and argument, faith and religion, and unscientific theories only; that the sanctity of all the many opinions on these subjects cannot be breached; and that to even *think* otherwise is to be a delusional know-it-all divorced from reality. As it turns out, the exact opposite is true.

Also as it turns out, all the secrets of the universe, at least all the ones worth knowing, aren't all that secret to begin with. Most have already been revealed to us and many of them for quite some time. The problem is that the world has refused to listen. It has not had "ears to hear." People believe what they are taught, and if what they have been taught is not the truth, so often that will keep the truth secret, hidden, and unavailable to them due to their own brainwashing and their own choosing. Therefore the pervading lack of understanding and confusion over the answers to the fundamental questions about this life is more a matter of willful ignorance on mankind's part, than any attempt by the *universe* to keep anything secret.

"For since the creation of the world the invisible qualities of God have been clearly seen, being understood through what has been made, even his eternal power and divine nature; so that men are without excuse." (Romans 1:20)

Follow the evidence

Science over the years has uncovered many secrets of the physical nature of this universe. By making careful observations and by painstakingly collecting facts and evidence, scientists were able to come to accurate conclusions, to reject erroneous assumptions, hypotheses and theories, and to unveil concrete scientific laws that still stand to this day. Laws of motion and of gravity, of gases and electricity, of light and of planetary orbits, and of energy and matter. And from these laws have come every modern invention and convenience that we enjoy, as well as take for granted today; our cars, and TV's, computers and cell phones, air and space travel, the marvels of modern medicine, etc. And all of these extraordinary advancements came about because these early scientists were able to reach accurate conclusions by observing the facts and evidence and by going wherever it led them. As Socrates instructs in Plato's *Republic,* "We must follow the evidence wherever it leads."

In the very same way, we can come to accurate conclusions and unveil correct answers concerning the questions of why we are here, what our purpose is, where we came from, where we are going, how and why this universe came into being, which ideologies are true and which are false, and which religious claims are true and which are false. By looking at the facts and the evidence, and going wherever it leads us, we can accurately answer these questions as well. Just as in science, we are interested in leaving the realm of opinion, conjecture, blind faith, speculation, false theories, false assumptions, false conclusions and false beliefs behind. And in turn we seek to come to conclusions and answers that are demonstrably true. That can be demonstrated to be accurate based on *all* of the facts and evidence, both historically and scientifically. We'll trust it to the judgment of you, the reader, whether or not this has been accomplished with each of the subjects we will cover. Again, people believe what they have already been taught, so if you disagree with any of the conclusions here just be certain that you *are* disagreeing based on the preponderance of the facts and evidence, and not because of any pre-existing bias, indoctrination or inaccurate assumptions within your own mind. If you seek the truth, you will find it. Conversely, if one seeks instead to protect the lies and indoctrination and false assumptions that they may now hold dear then, unfortunately, that's what they'll get. That's exactly what they will receive instead of the truth.

We are going to cover a number of subjects here and, despite the overall length of this tetralogy (4 books in one), we will still be limited in the amount of space we can devote to each one. The facts, evidence and logic that we will bring to bear in each of these areas, however compelling they will be, are still not the final word. There is a vast amount more that could also be presented to support these conclusions, and indeed at the end of each chapter there are numerous books and resources listed for this purpose. But it is not our intention nor do we have the space to exhaustively present all of the facts and evidence for each of these truths here. All we can do is expose the reader to these conclusions and hopefully enough of the overwhelming evidence that supports them. After that it is up to the individual to decide whether they have found the truth or not. But our attempt here is not to convince the inconvincible, as it is impossible to win an argument with an ignorant man. What we are attempting to do is to equip as many as possible with the truth so they are able to withstand the onslaught of lies from the dark side that have controlled our educational system, our entertainment industry and of course our media for some time now. And so they will have the ammunition to stand firm against those teachers, university professors, faux scientists, media charlatans, journalist imposters and late-night clowns who use their power, position and influence to lead the unsuspecting, the naïve, and the misinformed down their chosen path of confusion and lies; to keep them trapped on their mentally disturbed, far left, socialist, insane Venezuelan/ Cortez Democrat plantation. ☺

Decision time

There is a war being waged on this planet. It is for possession of the earth. It has been building for six thousand years and it will soon come to its conclusion. (See chapter 9.) It is playing itself out between the nations of men, the "sheep and goat nations," but the real battle has been and is going on behind the scenes, in the spiritual realm, between God and Satan. Now Satan, the god of this world, doesn't stand a chance. God will soon cast him out and retake complete possession of this planet. Indeed the only reason that he has allowed the battle to go on for this long is so another war that is being waged can also play itself out to its conclusion as well. And that war is also occurring behind-the-scenes, in the spiritual realm. It is a battle for the possession of the souls of men, and it has eternal consequences. It is being waged here on the inside, in everyone's heart and mind, and it involves an incredible gift that God gave to us; a gift that is both wonderful and terrible, depending upon how it is used; and that is the gift of our own free will. (It's one of the things that makes us in His image and His likeness.) While we are alive on this planet it is decision time. After we die that decision will stay with us for all eternity.

Yet as important as this decision is, most of the people on this planet have never been given the opportunity to look at all of the facts and evidence so they can make a wise and informed choice. Instead most of us have been deceived since birth. For the lies of Satan cover this earth like a blanket—false religions, false ideologies, false scientific theories—and people believe whatever they are taught. And most of us have not been taught the truth. This world is awash in a sea of lies that stem from one known as the Great Deceiver, and the people of this world are drowning in that sea. Satan's lies keep people from coming to the knowledge of the truth, and his lies are many. They are sly. They are pervasive. And their grip is powerful. As such we wanted to leave no stone unturned in exposing them, debunking them, refuting them, and in putting forth a powerful and indisputable case for the truth, so after reading this there could be no doubt or question in anyone's mind as to what the lies are, and to where the truth lies.

This is not a short story. It is a graduate course on the truth. But one that we feel is well worth taking, and passing. It is designed to take one from the lies of Satan (which cover this earth like a blanket) to the knowledge of the truth. It is designed to take one from darkness to the Light. Our goal here therefore is to give the reader the opportunity to do just that. To be able to look at all of the facts and evidence and to be able to do that within the pages of not hundreds of books, but just one work; and from there hopefully to make a wise and informed choice. To that end, good "luck," and may God richly bless you.

In conclusion then, it is our contention that despite what the current *wisdom* of this world might say, <u>the truth can be known</u>. The truth, not based on any man's opinion, but based on all of the facts and all of the evidence both historically and scientifically. A very clear understanding of the answers to the fundamental questions concerning this life *is* readily available to any and all who will seek it. It *is* possible to come to an understanding of all things and to uncover all of the secrets of this universe that are worth uncovering. You just have to seek, and you will find.

THE BOOK OF ETERNAL LIFE

(CONCLUSION TO ALL THE SECRETS)

"And I heard a great voice out of heaven saying, 'Behold, the tabernacle of God is with men, and he will dwell with them, and they shall be his people, and God himself shall be with them, and be their God. And God shall wipe away every tear from their eyes; and there shall be no more death, neither sorrow, nor crying, neither shall there be any more pain: for the former things have passed away.' And He who sat upon the throne said, 'Behold, I make all things new.' And he said unto me, 'Write, for these words are true and faithful.' And he said unto me, 'It is done! I am Alpha and Omega, the Beginning and the End. I will give unto him who thirsts of the fountain of the water of life freely. He who overcomes shall inherit all things; **and I will be his God, and he shall be my son**. But the fearful, and unbelieving, and the abominable, and murderers [and all of those who *vote* for the murderers], and whoremongers, and sorcerers, and idolaters, and all liars shall have their part in the lake which burns with fire and boiling sulfur, which is the second death.'" (Revelation 21:3-8)

In the book of Acts we read of the arrest and imprisonment of the Apostle Paul and his subsequent hearing before King Agrippa. Paul took some time to relate his miraculous and life-changing encounter with the risen Christ on the road to Damascus and his conversion from persecutor and murderer of Christians to a follower himself. "Then Agrippa said to Paul, 'You <u>almost</u> persuade me to become a Christian.'" (Acts 26:28) The operative word there is *almost*. And because of that *almost*, think for a moment of where that King has been for the last 2000 years, and the excruciating pain and suffering he has been heir to. I pray that no one who reads this book will be like King Agrippa, *almost* convinced of the truth.

Now you know the truth. "And you shall know the truth, and the truth shall make you free." (John 8:32) If you have ears to hear. "And He said unto them, he who has ears to hear, let him hear." (Mark 4:9) But for those who refuse to listen to the Truth, and instead turn their backs on Him, and continue to embrace the lies of the great deceiver, for them there is no hope. "And the devil who deceived them was cast into the lake of fire and boiling sulfur, where the beast and the false prophet are, and they shall be tormented day and night for ever and ever." (Revelation 20:10)

The lies of Satan do cover this earth and people believe what they are taught, and this world is full of people who desperately need at the very least to be given the opportunity to make a decision between what they have been taught, and the Truth. This has been the purpose and the work of countless numbers of brave true-Christian missionaries who have spread the gospel all around the globe for the last two thousand years. And it is for this ultimate purpose, to give people the opportunity to come from Satan's lies to the knowledge of the Truth, that this book was written. Peace be with you, and may God richly bless.

"I have no greater joy than to hear that my children walk in truth." (3 John 1:4)

* * *

"Then I saw a great white throne and Him who sat on it, from whose face the earth and the heaven fled away. And there was found no place for them. And I saw the dead, small and great, standing before God, and the books were opened. And another book was opened, which is the Book of Life. And the dead were judged according to their works, by the things which were written in the books. And the sea gave up the dead who were in it, and death and hell delivered up the dead who were in them. And they were judged, each one according to his works. Then death and hell were cast into the lake of fire. This is the second death. **And whosoever was not found written in the Book of Life was cast into the lake of fire**." (Revelation 20:11-15)

"I Jesus have sent my angel to testify unto you these things in the churches.
I am the Root and the Offspring of David,
and the Bright and Morning Star."
(Revelation 22:16)

ADDENDUM 2

ROCK THE TRUTH MINISTRIES

Rock the Truth Ministries was started in 2008 in direct response to the "Rock the Vote" movement which was very popular with the lying Democrat/ Left and their baby-killing media, and it was hilarious, and nauseating, how the self-righteous true believers actually thought they were doing a good thing. But in reality their movement was nothing more than an attempt by left-wing indoctrinated, baby-killing Democrat adults to try and get as many left-wing indoctrinated youths to get off of their blessed assurances and go out and vote for the baby-killing, lying, false-accusing Democrats. How wonderful. "Rock the Baby-killer Vote." Whereas Rock the Truth is the opposite of that. Our goal is to give as many left-wing indoctrinated youth, and adults, the accurate information necessary for them to be able to extract themselves from all of the political, as well as other, lies they have been spoon fed by a lying, baby-killing Democrat educational/indoctrination system, entertainment industry and media, in order to give them the ability to come to the knowledge of the truth so they can make an informed, rather than a brainwashed, decision as to whom to vote for and support. Which is the Democrats worst nightmare.

Here are some thoughts along those lines. They are posts from my blog, "It takes a Village Idiot. To Vote Democrat." (A take-off on Hillary Killery's book of a similar title.) Now you will find a lot of repetition because these posts were written individually over a few year period at times with the passage of weeks and months between them. But then isn't repetition the mother of learning? I hope you will find them informative, and maybe some might put a smile on your face. Feel free to copy, paste and share any of them on Fascistbook or any other social media, or anywhere you like if you think they might do any good. And keep in mind that the truth is very offensive, to those who are filled with lies, and who love those lies and automatically defend them no matter how stupid and idiotic they might be. ("ALL OF LIFE JUST MADE ITSELF!! SHAZAAM!!") Also keep in mind that what is true is based on all the facts and evidence. Anyone who is offended by the brutal truth here might want to think what actual facts and evidence they have to support their offense. Jus sayin...

I do want to mention two things before we get started. The first is that I am NOT a Republican. I am a follower of the truth. I am into the truth. Period. Based on all the facts and evidence. And it is the truth that destroys the Democrats and the satanic religion that they are trying to shove down the throats of decent America. And secondly, I want to make it clear that I do NOT condemn, demean, judge, put down, or disparage any girl or woman who has suffered through having an abortion. They have already been beaten up enough just by what they have gone through, and they don't need any further condemnation. They need the love of Christ which surpasses all knowledge, can heal any wound and forgive any sin. (Paul murdered Christians before his conversion, for goodness sake.) Indeed we are all sinners, some of us saved by grace, and repentance and forgiveness (in that order) is open to all. In addition, the only time I ever use the term *abortion* is when talking about those who have suffered through one, out of empathy for what they have experienced. For indeed, the very term "abortion" is the term of the enemy, Satan, and it is a lie. It is meant to whitewash and "put a happy face" on the horror that is actually occurring. No one is flying a jet in too low for a carrier landing: "Abort! Abort!" No. A baby is being mutilated to death while it is still alive. His arms are being ripped out of his shoulders and her eyes are being sucked out of their sockets; scissors are being jammed into the back of their skulls and their brains are being sucked out of their head with a f*****g vacuum. All while they silently scream to death. Also, many of those who have had an abortion are indeed victims themselves (along with their destroyed babies) of the lies constantly spewed by the fascist Democrat-Left that controls our educational system, media and entertainment industry: "It's your *right!*" "Why mess up your life?" "It's just a simple procedure." "It's your *choice.*" "It's not a baby anyhow." "It's just a lump of tissue." "It's only a *fish embryo* right now." My battle is with spiritual wickedness in high places, Satan and his demons who cover this earth, and his poor, unfortunate, Lake-of-Fire-bound suckers who, mostly out of ignorance, vote for and support his evil agenda.

POSTS

-What is happening in this country politically actually has nothing to do with politics but has everything to do with Satan's modern-day stealth, hidden and camouflaged religion of child sacrifice, bald faced lying, false accusations (i.e. against Kavanaugh, Trump and other actually decent, non-baby-killing people), child perversion, sexual and marriage perversion, America and Israel hatred, mutilating young girls and boys in the name of their new god- transgenderism, as well as other filthy evil things that his religion embraces. It is hidden and camouflaged because it is not recognized as one of his many false religions, but rather is seen as an "ideology," or a political movement, but it is a religion nonetheless. And the fact that his religion owns the Democrat political party, the Democrat media, and Democrat entertainment and educational systems means that Satan can spew his lies 24/7 and that he literally has complete control of the dissemination of information in our culture. We need to stop hiding behind the lie of "politics" as if that frees us from addressing Satan's lies and his control over his liars that own the Democrat Party and the Democratic media. We need to start witnessing to people along this heretofore taboo, so called "political" subject, so they can have a chance of freeing their mind from the lies of Satan and coming from darkness to the light. As Christians we are supposed to be here to bring people to the knowledge of the truth. All true Christians are supposed to be witnesses. And millions upon millions of people in this country are duped and suckered by Satan. On the one hand they actually think they're Christian. Cuz they go to church, regularly, and pray, and sing songs, and give tithes and offerings. Yet incredibly, on the other hand they vote for and support Satan's disgusting, filthy, child sacrificing, false accusing, hate-filled-violent fascist/antifa, totally INtolerant, baby mutilating, lying, etc. etc. modern day religion. And they are on their way to the Lake of Fire (see the Book of Revelation) because there will be no murderers, and no people who vote for and support murderers, in heaven. ("It would be better for you if a millstone were tied around your neck and you were cast into the sea, then you should offend millions of my precious little babies by voting to mutilate them to death.") ("Depart from me you workers of iniquity, I never knew you.") That's what actually is going on in this country. It's a battle between the truth of Jesus Christ, and the lies of his enemy Satan, the god of this world, the god that so many people are duped into serving with their votes and their support. I know the vast majority of people, both on the left and on the right, do not quite understand this. Yet. Unfortunately. They think it's "just politics." But it's far, far more than that. It's just eternity.

-It says in God's word that he will send them strong delusion that they should believe a lie. And the biggest lie that has been perpetrated by Satan the Great Deceiver, the Great deluder, that all of the elites believe, is called Satan's genesis. And it is meant to undermine the true account of creation found in the Book of Genesis in God's word. But know that it is a scientific fact that the theory of evolution along with its companion theory of the Big Bang are both so stupid, idiotic, asinine, ridiculous, absurd, and so scientifically impossible according to all of the facts and evidence that they are literally insane. Also conservatives Republicans and Christians should know that is Satan's genesis and it is the quicksand foundation of the baby mutilating lying, false accusing and America destroying Democrat left. If we fight this battle without destroying their foundation of evolution then we are literally flailing our arms into the air and beating nothing whatsoever. It is time to wake up Fox News, you poor uneducated evolution believers, along with the rest of America that does have a brain politically, and start understanding the weapon that we have in our hands to destroy the baby killing, lying Democrat left. And that weapon is the fact that their foundation, their theory of evolution and the Big Bang, are both so stupid and scientifically impossible that they are literally insane. I'm going to leave you with just one fact out of thousands upon thousands that totally destroy the theory of evolution. And that is the fact that you have 60,000 miles of blood vessels in your body and if you think, or any of the elites think, that those 60,000 miles just put them self in the exact right position throughout your body with far beyond micrometer accuracy so they touch every one of the 100 trillion cells in your body, and they did this accidentally by random chance, then you and they are literally insane. Forget about the fact that our blood system is under pressure. Think about it. How does a pipe system create itself and does so under pressure!? Yeah, you're right, that also is so stupid it is literally insane. What do they think? That as we "evolved" over hundreds of millions of fabricated years our blood system was at first of course NOT under pressure (in which case you're dead) until it "evolved" the necessary pressure?? (HahahahaHAH) And as far as the big bang and with all due respect to our deceased friend Mr. Hawking, if you are so stupid to think that an explosion created this universe then may I suggest you get yourself some dynamite, go into the woods and build yourself a housing development. You'll have the same chance that a universe has of creating itself by exploding.

-**"Look, the average Democrat voter is just plain stupid. They are easy to manipulate. That's the easy part."** Hillary Clinton as told to Dick Morris in 2005. Like I've been saying for many years, many people will watch the Democrat news media spew their idiotic lies and shout in frustration, "What? Do they think we're stupid?" And my answer always is, "No, they think their voters are stupid." They know their voters are stupid. And that's why they keep spewing those obvious lies. To keep them enslaved on their Democrat plantation.

-It's important to understand, if you want to get a good grasp of what's going on in this country, and on this planet, today, that Satan owns Facebook, Twitter, YouTube, Amazon, Google, Etc. as well as the Democrat Party, the Democrat media which is over 95% baby mutilating lying false accusing demon rat Democrats, along with that really particularly disgusting, clitoral-mutilating, worldwide primitive false religion that wants to shove it's sickness down the rest of the world's throat (in the name of their father the devil), as well as the educational system and the entertainment industry both of which are over 90% baby mutilating demon rats. He is the prince of the power of the airwaves. And keep in mind that a few thousand years ago God said that Satan is the prince of the power of the **air.** But I would expand that verse just a little, for today, "he is the prince of the power of the air**waves.**" Because a couple thousand years ago when that was written the readers did not understand TV, radio, the internet, Etc. and the Airwaves that produce that communication. But now we do, now we can see exactly what the Bible was talking about 2,000 years ago. And that it was written specifically for today. Beelzebub is the prince of the power of the air-waves. He controls, through his Demonrat, baby-killing followers, the TV, the Radio and the Internet. He is in control of spiritual wickedness in high places all across this Earth. And this is all a part of Prophecy... Satan and his Antichrist will come to rule over this Earth, hence the whole One World Government and globalist movement, for a terrible seven year tribulation period before the Lord returns and erases evil and child sacrifice and baby mutilation and lying and the filthy disgusting, satanic, women-hating, clitoral mutilators, and the false accusers from off the face of the Earth. But in order to do so, and to take over the world, Satan Must Destroy America, Must Destroy America's sovereignty and Power. And his left-wing dupes are doing just that.

-DESTROYING THE LIES OF SATAN... Satan is the great deceiver, and the god of this world. His lies cover this earth like a blanket, and people believe whatever they are taught, whatever they learn. False religions are his greatest lies... Roman Catholicism, Mormonism, Jehovah's FALSE Witnesses, Islam, Buddhism, Hinduism, Atheism, etc. as well as the false and wholly idiotic religions of our origins: Evolution and The Big Bang. Both of which are so stupid, so scientifically-impossible according to the Laws of science and all of the facts and evidence, and so ridiculous that they are literally insane. ("The Universe, and Life, just made themselves, accidentally, by random chance, over billions and billions of years." Yea, right. And the Easter Bunny built the Pyramids.) (By himself on a Sunday afternoon after lunch.) …They are Satan's genesis. And their purpose is to destroy the credibility of the true origins account in God's Word, and therefore to destroy the credibility of the rest of the Bible along with Christ and the Salvation He alone can offer.

But Satan's greatest and most effective false religion, in this country, is his baby-mutilating, lying, constitution-perverting, immoral, "progressive" (regressive), child-perverting and marriage-perverting Democrat/Left ideology religion. It is the religion of pure evil, that of Satan and his modern-day child sacrifice. And it wholly controls our media, our entertainment industry with its late-nite clowns, our educational system from elementary school through college and university, and of course his Democrat party, Satan's political/power arm to continue his policy of killing babies, perverting God's sacred institution of marriage, allowing boys who say they "are a girl" to shower naked with the girls in our public schools, etc. And through his religion and its control over all of these institutions, the great deceiver controls the dissemination of information in our culture. And keep in mind that he is an absolutely consummate LIAR. (The next time you listen to the "news" on CNN, or MSNBC, or CBSABCNBC, The Washington Compost, The NY Times…)

That's the false religion. But what of the true church?? …In 1st Corinthians 5 Paul admonishes the church in Corinth because they had sin in their midst and had failed to address it, but don't we have a far far greater sin in our midst with "Christians" going out religiously every two years and voting for, as well as supporting, and being enamored by the leadership of, this satanic religion of baby-mutilation and child-sacrifice?? Shouldn't we, as the church in Corinth did, address this iniquity? And therefore give those who are deceived by the great liar, a chance to rid their minds, and hearts, of this evil? "If anyone names the name of Christ let him depart from iniquity." And there is nothing more iniquitous on the face of this earth than mutilating countless millions of Jesus' precious babies to death while they are still alive in the "safety" of their own mother's womb, and going out and voting for the same, and supporting the same, and being enamored by the filthy, evil leadership of the same.

Also know that we do not put down, condemn, demean or judge any woman who has suffered through having an abortion. We are all sinners, some of us saved by grace. And repentance and forgiveness, in that order, is available to all. And this has nothing to do with "politics." It has everything to do with the Spirit, and with that same Holy Spirits battle with "spiritual wickedness in high places." It has everything to do with the Truth ("I am the truth"), and with that same Truths-Jesus' battle with the lies of the great deceiver. ...Is the church in America cursed with a curse? Because we have iniquity in our midst? And have done nothing to confront it? We pray for Holy Spirit revival, but America keeps plunging further and further down the cesspool tubes. May God have mercy on our souls. What have we done?? What have we NOT done?

-Here's a really incredible thing about the first verse of the Bible. "In the beginning God created the heavens and the Earth." A scientist a number of years ago became very famous because he was able to categorize everything in existence under five headings: time, being, action, space and matter. Everything in existence falls under one of these headings. And here's the really remarkable thing that they found out after that. The first verse in the Bible contains all five of the categories where everything in existence is covered under. *In the beginning*.... Time. *God*.... Being. *Created*.... Action. The *heavens*.... Space. And the *Earth*.... Matter. God is so incredible. He really blows you away sometimes. If you have ears to hear. And eyes to see.

-What's going on on this planet is a battle between truth and lies. On the one hand you have the truth of Jesus Christ, the Creator and sustainer of this universe and of life on Earth. On the other hand you have Satan the father of Lies, the Great Deceiver who goes forth to deceive the whole world. On the one hand you have the followers of Jesus Christ and his truth who love the truth and hate the lies of Satan. On the other hand you have those who actually love the lies of Satan that they were taught and who mostly unknowingly follow him and his lies. He is a child sacrificer of old and those people today who vote for the baby mutilating Democrats are one with him and his religion of child sacrifice and bald faced lying 24/7 through the media that he owns, and false accusing through his media, his entertainment industry, his educational system and his Democratic party all of which he also owns. This is not an argument between the right and the left. This is a battle between those who have come to the knowledge of the truth, and those who are still deceived and are on the side of Satan and his lies and his baby mutilation and his false accusations etc. etc. And those four million Americans who actually think they are Christians and on his side and yet march out religiously every two years and vote for Satan and his baby mutilating, lying, false accusing, filthy demon rat Democrats are terribly deceived. "It would be better for them if a millstone were tied around their necks and they were cast into the sea then they should mutilate millions of Jesus's precious little babies to death with their votes." "Depart from me you workers of iniquity. I never knew you"

-The gun control spastics were at it again at that farce "town hall" in Florida after the Parkland shootings. Unfortunately Dana Loesch just sat there and let those left-wing indoctrinated idiots shout her down with their false accusations of the NRA being "killers." This is what she should have said in response to their mindless assault… "What are you people talking about? The NRA are killers? Are you all really that stupid?? The real murderers are right here in this auditorium. Please stand up. Come on. That means all of you. All of you baby killers. All of you baby mutilating left-wing indoctrinated Democrat-voting and supporting murderers of the most innocent. How many babies have you mutilated to death to date? 60 million. Will the real murderers here please stand up. And it's not me or the NRA or any legal decent gun-owning Americans. It's all of you left-wing indoctrinated, mentally poisoned fools that believe all the lies you have been spoon-fed in our Democrap educational/indoctrination system. Your self-righteousness and false righteous is nauseating. You have the same righteousness as the Jew killers in Germany 80 years ago. Which is no righteousness and no morality whatsoever. You lie about the NRA and gun owners being the killers and being responsible for what happened at your school. When it is you and all of your fellow baby-killing Democrats who have destroyed the sanctity of life over the decades by slaughtering tens of millions of Jesus' most innocent and precious. That is the real cause. Shame on all of you. Now sit down and shut up."

-Someone just asked how evolutionists can explain the evolution of sexual organisms in every single species on the face of this Earth. Answer. … They can't! They just ignore the fact that the only way for sex to happen is that at the exact same time, on the exact same place on Earth both the male and the female would have to simultaneously evolve all of their sexual organs instantaneously, including all of the brain and hormone functions, nerve functions, smell attractions and visible attractiveness etc. etc. in order to make the entire sexual act and procreation possible. And that would have to happen in every species on the face of this Earth. Over and over and over and over again. Like I have said numerous times, evolution is for idiots. It's scientifically impossible. It is so stupid it is literally insane. Sex is just one of thousands upon thousands of facts and evidence that make a total laughing stock out of the theory of evolution.

-In case anyone cares to actually know the truth.... There is only one true religion and that is the Jewish Christian religion, which are one and the same. No matter how many Jews might still deny their own Messiah Jesus, and how many stupid false Christians might hate the Jews and Israel, like the jackasses of the BDS movement. The Jewish religion and the Christian religion are one and the same and it is the one true religion on Earth. It is also a fact that the Bible is absolutely written by God just like it says over 3,000 times. It is also a fact that Jesus Christ is God and he came here to Earth to die on a cross to pay the penalty for our sins, so if we so choose we can be covered in his righteousness when we die and therefore spend an eternity in heaven here on Earth. Think Hawaii on steroids. (As opposed to the foolhardy and stupid quest of attempting to get to Heaven covered in our own righteousness, which to God is nothing but "filthy rags"). Now that's the truth. You can do with it what you like. And that truth is based on all the facts and evidence. The Earth is round not flat. How do we know that? All the facts and evidence.

> "You are worthy, O Lord,
> To receive glory and honor and power,
> For You created all things,
> And by Your will they exist and were created."

And yet the jackasses give glory to stupid random chance. And shove that manure pile down the throats of our gullible and unsuspecting suckers in our elementary and high schools, colleges and universities. How pathetic. (What say you, Fox News??)

-The theory of evolution is an idiotic, asinine and scientifically impossible joke. How do we know that? All the facts and evidence. No matter how many stupid, idiotic, unscientific lies we were taught all of our lives growing up, and are still blasted with day in and day out from the left wing, lying Democrat media, educational system and entertainment industry. It is just another stupid, idiotic and asinine false religion. (And this world is full of those…) It is Satan's genesis, meant to undermine the historical accuracy and credibility of God's words and therefore to undermine the gift of eternal life paid for by Jesus Christ his Son. And all the morons in our national media, unfortunately including Fox News, spew that filthy false religion as if it is a "scientific fact." What a joke. (Hey Gutfeld, Tyrus, Hannity, Tucker, are you paying attention?) ....So there you go. Good luck and God bless

-The organization Black Lives Matter doesn't give a damn about black lives. They should change their name to Black Lives DON'T Matter. That would be far more accurate. Because if they cared about back lives they would be marching and protesting in front of the Democrats black baby exterminating, Planned Parenthood, baby mutilation clinics where 40% of all babies mutilated to death every year are black babies. Yet blacks are only 12% of the population. (Hey LeBron, need a calculator??) And that is by design. Miss Sanger, the founder of PP, said "We want to exterminate the blacks," and Hillary Killery said "Margaret Sanger is MY HERO!" BLDM would also go to Chicago and all the other democrat-run cities where blacks are getting killed every day by the truck loads. But they don't give a s**t about all of those deaths because they can't blame the cops. They would have to blame all the Democratic mayors that run those cities and have turned them into murderous s**tholes of black on black crime and murder. BLDM are nothing more than brainwashed, hate-filled, mentally deranged cop haters, racists themselves against their own race by being a part of the Democrat racist black baby exterminating party. They and their Democrats are the racists and not the police. Their self-righteousness as if they are better than the police and better than the rest of America is so disgusting that you could puke your guts out. But yet how the Democrat media just kisses their arses. Thank God that the truth is starting to come out by real black leaders and celebs and sports stars and police responding to the filthy lies of the FASCISTS of "Antifa" and BLM and the Democrats who own over 90% of the media, educational system and entertainment industry so they could spew their cop hating, anti-American lies 24/7 for decades without challenge. "There is no idea as powerful as one whose time has come." Maybe the time has come for the truth...

-And speaking of woman's suffrage, a friend once said that women fought really hard to win the right to vote. And then he asked, "Why?? So they could kill their own babies?" OUCH!! Time for some bodies to look in the mirror.

-The totally insane Demonrats have made a mockery of women's sports and pretty much destroyed it, while laughing at and spitting in the face of women athletes. "HEY!! LOOK AT ME!! I WAS BORN A MAN, FAR MORE POWERFUL AND QUICKER THAN WOMEN, BUT NOW

I THINK I'M A WOMEN SO I GET TO COMPETE IN WOMEN'S SPORTS!! AND WIN!! GOD, ARE THE DEMOCRAPS RETARDS OR WHAT???" Yes, they most certainly are. And women still will go out and vote for them. (See the previous post.)

The Word Nigger. (Definition: A contemptuous person. A stupid, ignorant and sub-human individual.)

Let's set the record straight. Black people are not niggers. They never were. It was the white, racist Democrats who tried to demean and disparage all black people by calling them niggers, that were the actual niggers, and not the black people they falsely accused. The really "stupid," "ignorant" and "sub-human" folks were not the blacks being falsely accused, but the stupid, ignorant, subhuman and racist white Democrats.

Again, black people are not niggers. Niggers are niggers. And they come in all shapes, sizes, colors, ethnicities, races and religions. Biden, that pathetic demented, senile, brain-dead, baby-killing, lying pig that never actually won the election, is a nigger. The Jew-killing Nazis from Germany 80 years ago were of course Niggers. I mean, they slaughtered Jews. Like those Hamas niggers did in Israel on October 7th. And in the same vein, the baby-killing, Nazi Democrats today in Congress, in the media (95% lying Democrat), the educational system and entertainment industry (both over 90 % Democrat) are also niggers. I mean, they slaughter babies. What about those Muslim Jihadi, innocent-people-slaughtering-bastards, like the inhuman pigs of Hamas and Hezbollah? Definitely subhuman niggers. And the real fascists of "Antifa"?? Niggers. How about all of those left-wing indoctrinated, baby-killing, Democrats that spew lie after lie from their positions as "journalists," "commentators" and anchors? Lie-filled niggers all doing the work of their father Satan who was a liar from the beginning and the father of lies. I could go on. But you get the point. And keep that in mind as you watch the "news."

-Recently Candace Owens was shouted down by an entirely brainwashed, ignorant TI at some event full of good little Democrap trained seals in the audience. TI is a useful idiot of Satan and is stuck on the Democrats racist, black baby exterminating, KKK Plantation, and he tried to mock both Candace and Trump supporters by asking "when was America ever great." Jackass. Here is my response to that brainwashed ignoramus.... Hey TI, America was great back in the sixties before blacks were suckered by Lyndon Baines Johnson into voluntarily being enslaved on his Democrat Plantation. After he said "when we pass this welfare reform we're going to get every Nig__r vote for the next 100 years." He was a racist. And his party was and is the party of racism. (Which you and LeBrainless James vote for and support.) ...Before Roe v Wade and black baby extermination became a reality by Planned Parenthood and the demon rat Democrat Party. Before Satan and his demon rats suckered blacks, with the help of the media, entertainment industry and educational system that they own, into spending an eternity in hell after they die because when you vote for murderers you are a murderer. And there will be no murderers in heaven. That's when America was great TI you stupid, ignorant, uneducated, brainwashed, useful idiot of Satan and his black baby exterminating Democrat left. And that goes for all of your brainwashed, useful idiot, trained Democrap seals in the audience. God bless Candace Owens. TI attempting to debate her is like an orangutan grunting at Einstein.

-Sorry. Wally World's closed. There is no evidence whatsoever for evolution or the Big Bang. They are both so stupid ignorant asinine ridiculous and so scientifically impossible according to the laws of Science and all the facts and evidence that they are both literally insane. You have 60 thousand miles of blood vessels in your body. In just one human body. How exactly did all that put itself in the exact location with far beyond micrometer accuracy so it touches every cell (there are a hundred trillion of them) in the body, just accidentally by random chance? Answer? It didn't! That fact alone, which is only one of thousands of facts, destroys Evolution. It doesn't exist. I just destroyed it. With three words, sixty thousand miles. The Earth and the universe are actually around 6,000 years old. And that is a scientific fact according to all of the evidence. The LIE of hundreds of millions and billions of years was merely concocted in order to give the false theory of evolution the miracle of time that it needed to create all of this. Of course if you gave evolution an infinite number of years it couldn't create even one DNA molecule, much less a "simple" single cell, according to the laws of statistics and probability. ("Simple"??!!@#%@?? One cell in your body is actually more complex than the entire city of New York.) Total horse shit is what we've all been taught...

-Atlanta Mayor hosts drag queen Story Hour for children at City Hall. ...... "Burning Alive in Hell. Burning Alive in Hell. Hi ho the Dario, Burning Alive in hell." Come on everybody! Sing along with me! "Burning alive in Hell. Burning alive and hell. Hi ho the Dario, Burning Alive in hell." Sung to the tune of, The Farmer in the Dell. ...What a sick bastard. And who votes for a subhuman pig like this?? (OH. That's right. Democrats.) ..."It would be better for you if a millstone were tied around your neck and you were cast into the sea, then you should mentally and spiritually poison even one of my little ones with your filthy sexual perversions." "Burning alive in Hell. Burning alive in hell. Hi ho the Dario. Burning Alive in hell."

-Here's a message from God to our Democrat friends on the other side of the aisle. This is from Dan's Amplified version.... ..."For the wrath of God is revealed from Heaven against all ungodliness and unrighteousness of men, who hold the truth in unrighteousness. Because that which may be known of God is Manifest in them, for God has showed it unto them. For the invisible things of him from the creation of the world are clearly seen, being understood by the things that are made, even his eternal power and all-knowing Mind, so that they are without excuse. [The evolution believing, God hating, baby mutilating, child sacrificing, Satan worshipping demon rats have no excuse. Because you have to be an absolute idiot filled with willful ignorance to believe the fake origins slop that they believe, and then teach our kids.] ...Because that, when they knew God, they glorified him not as God, neither were thankful, but became vain in their imaginations, and their foolish heart was darkened. Professing themselves to be wise, they became fools, [They think they are so smart believing that life just made itself, when at the same time if you ask them they'll tell you that of course a simple paperclip could never make itself. Idiots. And they think themselves so self-righteous and morally superior to everybody else (like the Ns of The View) when they practice the filth of baby mutilation. They have no morality or righteousness whatsoever. They are one with the Jew killing Nazis of Germany 80 years ago. Yet they own our media, our educational system and our entertainment industry so they can spew their disgusting false righteousness 24/7 without any opposition, like farts and diarrhea out of the baboon's ass stuck to the front of their faces].. ...They changed the truth of God. [the obvious fact of creation as spelled out in the Book of Genesis] ...into a lie [evolution and the biggee explosion] and worshiped and served themselves and their own utter stupidity more than the Creator, who is blessed forever. Amen. For this reason God gave them up unto vile affections, and even as they did not like to retain God in their knowledge, God gave them over to a reprobate mind [a mind that is literally insane] to do those things which are filthy, vile and satanic. Being filled with all unrighteousness, fornication, wickedness, covetousness, maliciousness, full of envy, child murder, deceit, malignity. Whisperers, backbiters, baby killers, haters of God, despiteful, proud, boasters, inventors of evil things [like child gender mutilation], pathetic fascists of 'antifa', disobedient to parents. Without understanding, stupid, covenant breakers, without natural affection, unmerciful. Who knowing the Judgment of God, that they who commit such things are worthy of eternal death burning alive in the molten lava of hell, not only do the same, but have pleasure in them that do them." Romans 1. ....The modern democratic party, full of insanity and filth, pure evil and disgust. And the brain-dead, brainwashed, uneducated idiots who listen to them and go out and vote for them. Maybe somebody ought to try and wake them up? Before they die and are thrown into the Lake of Fire (Sulphur stinking and boiling and burning at 1,000 degrees) forever and ever. Their reward for worshipping Satan and loving his lies…

Hey! Have you all heard?? They just renamed Critical Race Theory! It's now called, Crack of Your Ass Theory. Which is of course a much more appropriate name for this stinky, disgusting, racist pile of horseshit lie/"theory" that has nothing to do with truth or reality. But it is a really neat way for the stinky, disgusting, baby-mutilating, black-baby-exterminating, false accusing, racist Democrats, who own 100% of our social media, 95% of our national media and 90% of the educational system, to put forth their filthy idiotic scam designed to divide America and falsely accuse white people of being racist. Congratulations you satanic, lying bastards.

-It is a scientific fact that the theory of evolution is so stupid, idiotic, ridiculous, asinine, absurd and so scientifically impossible according to the laws of Science and all the facts and evidence, that it is literally insane. Yet how many fools spew the ridiculous lie, "the theory of evolution is settled science"? (Yeah, and my ass is the third Reincarnation of Buddha.) …The only question that remains America, is why you are so stupid, ignorant and uneducated to keep shoving this pile of non-scientific horseshit down the throats of our gullible, naive and unsuspecting school children in our colleges, universities, elementary and high schools. What level of moron does it take to actually believe that life could just make itself? If you show those same people a paperclip and ask them where it came from they will say without batting an eye that somebody made it. Yet one cell in your body, and there are a hundred trillion of them there, but that one cell is literally more complicated and meticulously designed than the entire city of New York. The cell has tens of thousands of pre-programmed nanobots running all throughout it from the cell wall and back to certain specific parts of the cell to carry oxygen and carbohydrates etc. and then take the waste products from that specific area of the cell back to the cell wall for elimination into the bloodstream. The DNA computer at the heart of each cell has more coded information than a small Library. Could a book just write itself? Also it is a scientific fact that coded information can never ever happen on its own accidentally by random chance. The odds of that are way too astronomical to ever happen. But the same evolutionary indoctrinated and brainwashed geniuses who are so smart that they know of course that a paperclip could never ever make itself, will look at life and say Shazam! It all just made itself. (Are you listening, Fox News?) Now I know people believe whatever they are taught, and I believed it too cuz I was taught it as well, so we're not trying to put anybody down here or judge anybody as if we are better than them. But to turn Richard Dawkins famous quote around on himself, "what level of knuckle-dragging ignorance does it take to believe" that life just made itself? Somehow. Miraculously. It didn't of course. It never could have. It was created. Just like your paper clip. Shazam!

-The demon-possessed rats are baby killers. That says it all. That's all you need to know. Would you spend any time debating the Nazi Jew killers in Germany 80 years ago? Of course not. You're not that stupid. Well somebody explain to me what's the difference between the Nazi Jew killers in Germany 80 years ago who slaughtered six million Jewish men women and little children, and the Nazi demon-possessed-rat Democrats in America who have mutilated 60 million babies to death, and are still going strong, and are so proud of it? There is none.

-You know the really incredible and incredibly sad thing in America is how many millions of democrat voters there are who actually think they are Christian and followers of Jesus. Somebody needs to wake them up big-time, before they die. Most of them have no idea of the pure filth they go out and vote for cuz they are deluded and deceived by The Great Deceiver, Satan who owns the Democrats and their media, educational system and entertainment industry, who all spew his lies into their brains 24/7. Satan and his Democrats espouse a filthy religion of child sacrifice, lying 24/7 to keep people trapped in this disgusting religion, false accusations (which the Lord hates) against their opponents (Kavanaugh, Trump, true Christians, Conservatives, Republicans), child, sexual and marriage perversion, Christian persecution, America hatred and Israel hatred. ...Here's the Democrat version of Christianity: Mutilate babies to death with your votes, love Jesus. Falsely accuse Trump viciously, love Jesus. Parade transgender perversion in front of our little children in first grade in our Public Schools, "love" Jesus. Spit in God's face with homosexual marriage, love Jesus. Be a part of spewing the lies of Satan 24/7 through the Democrat owned media, educational system and entertainment industry, love Jesus. Be totally bamboozled into voting for the filthy, racist Democrats and their slow genocide of the black race by their planned perpetration of black baby extermination, "love" Jesus. Get it? No, I'm afraid they don't. May God have mercy on their souls. ("It would be better for you if a millstone were tied around your neck and you were cast into the sea, then you should offend millions of my precious little ones by voting to mutilate them all to death." ..."Depart from me you workers of iniquity I never knew you.")

-Just saw a post about how a college put up signs around its campus saying HATE IS NOT WELCOME HERE, because the Democrats KKK were having a rally nearby. Well if hate is not actually welcome here, you stupid Democrats, then you ought to send a message to yourselves. Our universities and colleges, the Democrat Party, our entertainment industry and our media are full of nothing but hate towards Donald Trump and his voters and supporters* who actually have a brain, and for no reason except that they lost (thank you Jesus), and they hate his policies because, unlike theirs, they are not meant to destroy America. Yeah, "hate is not welcome here," you left-wing indoctrinating hypocrites. *(And now in Nov/Dec 2023 we see their filthy hate toward all Jews and the nation of Israel! And yet Jews go out and vote for them in droves!%@$@%!!??)

The Weather Channel… Man-made global warming jackasses spewing their lies 24/7.

-How in God's name can any black ever vote Democrat?? ....cuz roughly 40% of all babies mutilated to death every year by the demon rat Democrats and their racist Planned Parenthood are black babies. Yet blacks are only 12% of the population. Do we need a calculator Lebron? It's called the slow extermination of the black race. And it's by design. Cuz Margaret Sanger, the hero and founder and god of Planned Parenthood back in the thirties when she started it said "we want to exterminate the blacks." Just like her friend Hitler over in Germany was exterminating the Jews, she was doing the same thing over here. Except the blacks were her target cuz she thought they were inferior and she was trying to help the human race along. You know, the LIE of evolution. Survival of the fittest. And what did Hillary Clinton say recently? "Margaret Sanger is my hero." Black voters would know that if they listened to Candace Owens and other real black leaders. You know, the kind that are not allowed on the Nazi Democrat, lying, satanic, racist, black baby Exterminating Demonrat news media. Also, most Planned Parenthood abortion mills are in poor black neighborhoods. Why? That's not where the money is. No, that's where blacks are and Margaret Sanger is dead but she is fulfilling her dream of exterminating the blacks from the grave. Now that's the truth. Pass it on. Cuz blacks have been voting for decades for the slow extermination of their own race. It's pathetic. But that's what happens when you're lied to 24/7. That's all you're going to know. And to keep people from ever hearing the other side, what is the devil's biggest lie of the last 20 years?? 3 words. "Fox News Lies." Fox News is the only outlet that does NOT lie. But Satan needs to keep people trapped on his black-baby-exterminating, racist Democrat plantation so he can suck them into hell after they die. There will be no murderers in heaven. Nor those who vote for the same… Also, let's talk about who the real racists are. Trump?? What a hoot. No, the Democratic Party is now and has always been the party of blatant racism. They still hold blacks in complete contempt. In addition to their black baby extermination program going on for decades now, as far back as the Civil War they were the party of the South that wanted to keep blacks enslaved. After they lost the war they were the party of the Ku Klux Klan that wanted to stop blacks from voting so they hung them if they tried to, and they hung Republicans who tried to register blacks to vote, and they hung them just for the hell of it to keep them full of fear. They are the party of Jim Crow laws, and the utter destruction of the black family. I could go on but I think that's enough information for now. Good luck and God bless.

The hoax of manmade global warming has been perpetrated fanatically by the brainwashed, intellectual idiots that control our media, our educational system, and our entertainment industry. To the point where they have turned millions of their naive and unsuspecting youthful, useful targets into spastic, hysterical jackasses whose minds are absolutely infested with that obvious lie. The sun controls the heating and cooling of this planet. And man is absolutely inconsequential. And the fact is even if man does raise the CO2 levels of this planet then all it's going to do is lead to another Green Revolution that we have had numerous times in the past when that did occur without any help of man. Wake up you stupid, brainwashed, spastic, hysterical suckers. With all due respect of course.

-Friends don't let friends vote Democrat. Because if they do they will also be letting them spend an eternity burning alive and screaming at the top of their lungs in the Lake of Fire. No matter how good they think, and you think, they are. For there will be no murderers in heaven. And no one who votes for them. "It would be better for you if a millstone were tied around your head and you were cast into the sea then you should offend millions of my precious little babies by mutilating them to death with your votes." "Depart from me you workers of iniquity I never knew you."

-How can a Jew even be a Democrat for God's sake? They are Jew-haters and Israel haters. In addition to perpetrating a modern-day Holocaust that has mutilated 60 million babies to death. After the Holocaust that occurred in Germany 80 years ago you'd have figured they'd be a little sensitive to being a part of another Holocaust. And don't try to accuse me of being anti-Semitic. There is no stronger supporter of Israel and the Jews than myself. This is the truth. And the truth is not "anti-Semitic," or anti anything, except for being anti lies of Satan. And it is those lies that keep people trapped on his modern day Democrat Plantation. Wake up! Jews and Gentiles. Blacks and whites. Get off!

-How many here have been censored for 30 days by the Nazis who control Mark Suckdickerberg's Fascistbook? I have. Dozens of times. For posting the truth, which of course violates their "community standards." (The kind of "standards" that ooze up from the cesspool of hell.) Which means I'm not a mentally poisoned, left wing indoctrinated, child sacrificing, baby mutilating, racist, black baby exterminating, bald faced lying, filthy false accusing, America hating, Christian persecuting, Israel hating, child perverted, sexually and marriage perverted, demon-possessed-rat, Democrap proselyte of Satan's modern day, stealth, hidden, disguised and camouflaged religion of all of the above. Like they are. ☺ ....You know sometimes I feel like I fell into a wormhole, like a Time Warp, and ended up in 1930s Nazi Germany. Surrounded by Jew killers. Who own the media, the entertainment industry, and the educational system, and control the dissemination of information in the entire country, and censor anyone who's not a Jew killer like them. Has anybody else felt this way?

-We live in an insane asylum, and the inmates have taken over, ...our media to the tune of 95% of it, and over 90% of our entertainment industry and educational system. The country has been overrun by piles of pig vomit. Now let's all go out like a bunch of uneducated brainwashed yahoos and vote for them. Adam Schiff4brains has a baboon's ass stuck to the front of his face. Out of it he spews lies farts and diarrhea constantly 24/7 all over the poor dumb fools who are stupid enough to actually listen to him. (With all due respect of course.)

-Warning to those who believe in the idiotic and scientifically impossible lie that is called the theory of evolution. What you are about to hear is the truth. And the truth can be very offensive sometimes, to people who have been taught lies. So be forewarned.... Question, what level of moron does it take to think that life just made itself, accidentally by random chance? What level of moron does it take to think that the muscular system of the body, astronomically complex and perfectly placed in thousands of different variations throughout the body, just made itself accidentally? What level of moron does it take to think that your spinal column with all of its absolutely perfectly and minutely detailed and designed vertebrae with the multitudinous nerve fibers of the spinal cord running perfectly through it, just made itself accidentally by random chance? And your face just made itself, accidentally by random chance? If your face just made itself accidentally then why isn't your nose in the middle of your ass? Why isn't your mouth on the bottom of your foot? I'm sorry, but the theory of evolution is so stupid, so laughably ignorant, and so absolutely scientifically impossible that it is literally insane. Hey Greg Gutfeld, are you listening? Tyrus, are you paying attention? And keep in mind, kids, that it is the theory of evolution, Satan's genesis, which is the quicksand foundation of the baby mutilating, lying, false accusing, satanic, demon rat, Democrat left. If you're fighting their lies but you're too stupid to pull the rug out from under their biggest lie of all, then you are doing nothing more than blindly flailing away at them with your penis. Note to Fox News, conservatives and true Christians in the media and in the Republican Party, you are fighting an enemy who is a liar. He has a name. It is Beelzebub. You fight the battle against him by exposing his lies when they are spewed by his followers, but you are too stupid to expose his greatest lie of all, evoluton. Why is that?

-"Reform the police! Reform the police!" If I hear another word about how we need to reform the police, I'm going to puke. What needs to be reformed is the Democrats welfare that has destroyed the black family (by design) and flooded our cities with out-of-control, mentally-poisoned, left-wing-indoctrinated, violant thugs that are killing each other. THAT'S the problem. And NOT the police.

-I think there is something that really needs to be addressed within the so-called Christian churches of America. In First Corinthians Paul exhorted the church at Corinth to address the sin that they had in their midst. And they did so, in love, and the guilty party repented and was able to remain in fellowship with the church. What we need in this country is the first letter to the Americans. Because we have sin in the midst of our churches and nobody addresses it. Instead they just let people blindly march out religiously every 2 years and be a part of Satan's modern-day stealth religion of child sacrifice by voting for the baby mutilating, lying, false accusing, sexually-perverted Democrats. How sad. And that's not political. That's spiritual. It's a satanic religion, and Christians should have no part in it whatsoever. But they've been lied to all their lives because Satan owns the media, the educational system and the entertainment industry. Somebody needs to explain the lies in their heads so they can truly follow Jesus Christ. Before they die and are in for a rude awakening. "Depart from me you workers of iniquity, I never knew you." And there is nothing more iniquitous on the face of this earth than voting to kill babies. Which is a thousand times worse than a woman or girl who has an abortion. She is under great stress and anxiety, most of the time, whereas the voters are not. They are just stupid. And uneducated. Lied to. And that is the only time I will use the term "abortion." Out of respect and empathy for what the victims of killing their own babies have gone through. Indeed, "abortion" is the term of the enemy and his followers. It is meant to put a happy face on his evil "procedure." No one's coming in too low for a carrier landing. "ABORT!! ABORT!!" No, a baby is getting his arms yanked off, her eyes sucked out of their sockets, mutilated to death while they are still alive and screaming in pain. …Also, know that I do NOT judge, demean, vilify or condemn any woman or girl who has had an abortion. We are all sinners. Some of us are saved. By grace. And most girls have been heavily lied to all of their lives in the baby-killers "educational" indoctrination system. "It's just a "lump of tissue." It's "not human." It's just a "fish embryo." It's your "right" to kill it. (Do rights come from Satan?) No, my battle is with "spiritual wickedness in high places." With Satan, the prince of the power of the airwaves, and his many demons that cover this earth. And his many stupid human followers. Amen and amen

-The Democrats own the media, the educational system and the entertainment industry. As such they and their father Satan own the dissemination of information in our culture. In case anyone is wondering why the Democrats don't get investigated, don't get indicted, don't go to jail, and get away with spewing lie after lie after false investigation after false indictment against our former president who actually DID get elected. And that is also why poor unfortunate suckers march out religiously every two years and vote for them. They might know the pure evil that they support if they stopped listening to CNN. (And MSNBC and PBS and ABC, NBC, CBS, the Weather Channel, etc etc.)

-I would love it if I could just drive by a church in America one time that actually had the guts to have this posted on their sign by the road… "ALL ARE WELCOME HERE! BUT KNOW THAT THIS IS A DEMOCRAT FREE ZONE. WE DON'T KILL JESUS'S BABIES. NOR DO WE VOTE FOR THOSE WHO DO. GOD BLESS."

-If anybody watched that miniseries on the Clinton scandals and relived what the Democrats in the media and entertainment industry and the late-night clowns did to "that woman" Monica Lewinsky it should make you sick. It makes a mockery of their #metoo movement. Like they care about women now. They ruined her to the point of near suicide, and just so they could defend their baby killing, rapist (see: Juanita Broderick), Clinton. That's like the Nazis all of a sudden rising up and saying we care about Jews now, vote for us you stupid retards cuz now we are the party for Jews!! We are the #meJew movement now. And hilarious #mestupid women vote for them.

-Nancy Pileofshitsi is replacing the sign, IN GOD WE TRUST, in Congress above the door to the chamber, to the more Democrap appropriate one, SATAN'S ASS WE LICK.

"You smear somebody with falsehoods and all the rest. And then you merchandize it to the press. And then once the Press quotes it you sit there and say 'See, it's all true cuz it's in the Press.'" That's a quote from Nazi Nancy. Like I've said, the demon rat Democrats are all professional liars, adhering to the Nazis famous Mantra, "if you shout a lie loud enough and long enough, enough stupid people will believe you". They spew their obvious lies 24/7 in order to keep their stupid voters enslaved on their Democrat Plantation.

-The air is absolutely ever present everywhere. Anywhere you go on Earth. To the depths of the ocean there is air. To the tops of the mountains there is air. Go down as deep as you want in the earth and air is there as well. It is inside of you and saturates every inch of your body, and mind, your blood system and in all of your cells. Air is ever present everywhere. Think about it. And now understand that God is even more ever present everywhere than the air.

-Folks are saying that we need to expose how much of a liar Joe Biden is. And of course he is. It's been documented many, many times over. And that's all just terrific. …But here's a thought, maybe we should expose how much of a baby killer he is. ….Oh, that's right. Everybody already knows that.

He's a baby killer like all Democrats, and their voters, are. ... But nobody can call him that. Because Satan seems to have a control over everybody in this country that has a voice whether that's Republican politicians or conservative media types or even Christian pastors and leaders who get on TV. Nobody ever speaks the truth. They are baby killers. (Just like the Nazi's in Germany 80 years ago were JEW KILLERS.) That's who the Demonrats are. BABY KILLERS. But nobody has the balls to actually say it. It's sickening. You can't win a war unless you identify your enemy. Everybody screamed about that during the Muslim Obama's years when he refused to call the Muslim terrorists what they were, terrorists. But we live in a country where all of the idiots on the right side can't even identify who our enemy is. They are baby killers. That's what they are. Say it. It's never said. Say it. Nationally. Over and over again. Say it. 24/7. Say it. Scream it. Shout it. The Democrats are baby killers. Say it. Or we have lost the war.

-The ancient Israelites fell so far into demon worship that they would burn their kids alive as a sacrifice to Baal (Satan) on the outskirts of Jerusalem. The place was called Gehenna which soon turned into a synonym for hell itself. Loud drums were played to drown out the cries of children being sacrificed in the demons fire. "The more things change the more they remain the same." Kind of reminds you of the Democrats today, doesn't it?

-If you were to listen to the moron Adam Schifforbrains drone on spewing lies, farts and diarrhea out of the baboon's asshole stuck to the front of his face, you might find your crying out in frustration, "WHAT? Does he think we're stupid? WHAT, does he think we're THAT stupid?" And the answer of course is no, he doesn't think we are that stupid. He thinks his voters are that stupid. In fact he knows his voters are that stupid and that is the reason for this whole coup d'état attempt. This whole idiotic lie of a false impeachment is for Democrat morons, in Congress, in the media, in our educational system and entertainment industry and of course in their voting public. Yes, they really are that stupid. With all due respect of course.

-Satan needs to destroy America. He has to because we are the great protector of Israel, except for the eight years of that disgusting America- and Jew- and Israel- hating Muslim that the fools did what they were told to do and went out and voted for. And it is Israel ultimately that Satan has to destroy before the Lord returns and he is thrown into the Lake of Fire for a thousand years as prophecy details. (See the Book of Revelation.) He won't succeed in his demonic quest, but that's his goal that he is striving for like a maniac.

And he whips his faithful followers: his baby mutilating lying false-accusing Demonrats, his BDS morons, and his clitoral mutilating, genocidal head-choppers, into a Trump-deranged frenzy to accomplish just that.

This is good versus evil. And the Democrats are evil. In case anyone is too stupid to figure that out. Like Democrat voters for example.

Did you hear the news? (December 2023) Biden has decided to choose the leader of Hamas as his Vice Presidential running mate. …He's trying to win back the Muslim vote.

-The NBA. AKA The National Bitches Association. This ignorant, left wing indoctrinated troop of stupid rich "woke" hypocrites recently put an end to any support for the freedom fighters of Hong Kong, just so they can further pad their inflated financial pockets by licking the ass of the racist, oppressive communist state of China. Steve Kerr (the "phony, cowardly little corporate stooge that you are"… as Tucker Carlson remarked), LeBrainless James, Seth Curry, James Hardin (that clown actually apologized to China because his teams GM tweeted support for the Hong Kong protesters), and the rest of these hypocrites have just made America sick to its stomach. Congratulations "champions." America supports real men and women putting their lives on the line for freedom in Hong Kong and elsewhere, and not hypocritical cowards like the Nat'l Bitches Assoc. doing the bidding of their communist masters in Beijing. Now go out and vote for and support the slow genocide of your own race. Jackasses.

"It's the end of the world as we know it. And I feel fine."

God is absolutely sovereign. He controls everything, in fact he wrote it from before the foundation of the world which was about 6,000 years ago, and not hundreds of millions and billions like the evolutionist and Big bang theorist have concocted. He controls the movement of every leaf on every tree across the entire globe at every millisecond in time. Nothing is by chance. But at the same time we have absolute free will. If He wasn't absolutely sovereign then He wouldn't be God. And if we didn't have absolute free choice then He wouldn't be a just God. So as such he absolutely controls everything happening in the affairs of man. In fact it's already been written. It's His story. History. From prophecy we know where we are heading ultimately and that is a One World Government with the satanic globalists taking over, which of course can only happen with the destruction of America's sovereignty.

And then Satan and his Antichrist will rule over the world and there will be a seven-year tribulation period of horror. And then the Lord will return and wipe from the face of the Earth all evil, all the baby mutilators, all the Demonrat liars who own the media, all of the filthy false accusing demon rat Democrat politicians, and of course all of the Satan worshipping clitoral mutilators. In the end there will be perfect justice.

-TRAFFICKING… How much money, and therefore power, has a particular Government Intel Assoc. (GIA) accumulated over the years and over the decades by kidnapping children all over the country and all over the world and selling them to very rich pedophiles and sexual perverts all across the earth? How much day in and day out lifelong pain, emotional torment and horror have they visited upon tens of thousands of parents and siblings and friends and relatives? Child trafficking is a very large and lucrative$$ business$$. How many high profile pedophiles are there in Hollywood (Tom Hanks?) and in the leadership of the satanic Democrat party? (DC Pizza Parlor??) And how many actors and others who have threatened to expose them have been wasted by the GIA to protect them? How many politicians have been blackmailed by being secretly picked up by the GIA and forced to have sex, while being filmed, with one of their very young trafficked victims? And threatened with the torture and death of them and their wives and children if they refuse? (Their Democrat media never reports on it.) How many adrenochrome drinkers are there in Hollywood and DC who get this pure evil health and youth elixir drug from the blood of tortured little children and babies? (Know that all of them are already burning alive in the Lake of Fire forever and ever.) Does the GIA sell it to them? Is this one of the reasons why all those baby-killing, satanic elites hate President Trump so much? How many people who have gotten a little too close to, and learned a little too much about Hillary Killery, have died or disappeared mysteriously? There's a long list. Was Joan Rivers taken out because she threatened to expose that Michael Obama is a man? ,,,So many questions. So little time.

-"The heavens declare the glory of God, the skies proclaim the work of his hands. Day after day they pour forth speech, night after night they display knowledge. There is no speech or language where their voice is not heard. Their voice goes out into all the Earth, their words to the end of the world" Psalm 19: 1-3. .... What God is saying is that those who believe and teach the idiotic, scientifically impossible theories of evolution and the big explosion are deaf. They are blind too, but the above is just letting us all know how deaf they are. And dumb. And blind. 😁

-The truth about school shootings... 9/11 was a warning to America. A warning to the baby mutilating Democrats, and to all their brainwashed, uneducated voters, to stop killing Jesus's precious babies. But of course they didn't listen then, and they're not listening now. Nor is the rest of America sharp enough to know what's going on and to warn these demon rats of just the kind of wrath they, and all the rest of us, are going to shortly incur. You aint seen nothin yet America. So are the mass shootings in Midland and Odessa Texas, as well as the ones that happened in El Paso and Dayton. They are horrific and satanically inspired. But equally disgusting and satanic is the lying, false accusing Democrat hypocrites in the media, in Hollywood and in the Democrat congress, spewing their fake outrage and their fake horror when each and every mass shooting occurs, when in reality they are responsible. Not only that (as we will see shortly) but they actually rejoice inside because it gives them another opportunity to falsely accuse and point their rotten fingers at the NRA, and Republicans, and Trump, and the Second Amendment, when they should be looking in the mirror. Because it is those baby-killing bastards that have destroyed the sanctity of life over the decades in this country. They have cheapened life dramatically in the eyes of our young people growing up in their Demonrat run educational/indoctrination system where they are told to go ahead and fuck away, use a condom maybe, but either way go ahead and just do it. Cuz if you get pregnant you can just kill the thing. And they continue to get away with spewing their lies and diverting the attention from themselves every time a shooting occurs because the Republicans, and conservatives, and right-wing commentators, and Fox News are too stupid or too cowardly to point out the truth that we just did here. I guess there's some unwritten law that the Right has signed onto which says you can't call the Demonrat baby killers what they are, baby killers. And keep in mind that in the same time that those 29 people were shot to death, 3,000 innocent precious little babies were mutilated to death. And the sanctity of life was cheapened even further and the road to Hell on earth here continued to be paved even more, for more of these horrible shootings to occur in the very near future, and over and over again, until America rises up and demands that all life is sacred. That's the truth Republicans. Do what you want with it. [And as a footnote, and if I can get real with everybody, ask yourself for a moment how God the Creator of the universe the all-powerful, all-knowing ruler of Heaven and Earth -who also allows man to do his own thing with his free choice - but how do you think he feels about, just in America here, about us mutilating 60 million of his babies to death. What do you think he's going to do? Throw a party? In heaven? Go on a family trip to Jupiter to celebrate? No stupid Democrats, what you are getting in these mass shootings as well as many many other things that are occurring in this country and are soon to get even far worse, is you're seeing God's gift in return bakatcha for what you and your father Satan have gifted to Him, over these many terrible and genocidal years. "Woe unto that country that murders its way into the hands of an angry GOD." Good luck and God bless]

-The Jew killing Nazis from 80 years ago in Germany are alive and well and have taken over the baby killing Democrat Party, media, educational system and entertainment industry...

-"GLOBAL WARMING" HORSESHIT SCAM. Little emotionally distraught Thunberg, with her orchestrated spaz attack show for the sick retards at the UN, is mentally poisoned and mentally deranged with the lies of her parents and her teachers who are global warming insane extremist socialist fanatics who, like their soulmates the Jew killing Nazis in Germany 80 years ago, have no righteousness or morality whatsoever. But these hysterical, utterly brainwashed tards spew their lies with nauseating self-righteousness and false righteousness loudly and incessantly like farts and diarrhea spewing out of the baboon's ass hole stuck to the front of their faces. They make real people sick to their stomachs.

-The second law of Thermodynamics clearly states that nothing in this universe goes from a state of disorder to a state of order, unless acted upon by some outside intelligent Force. To say that some explosion many billions of mythological years ago created the universe spits in the face of the second law of Thermodynamics. No theory that is introduced that spits in the face of an already established law is even considered and can't even be printed in a scientific journal. It and its author are laughed at by the scientific community. But Big Bang Theory like its companion Evolution Theory are religions so they get a pass. ...According to the laws of mathematics statistics and probability nothing with odds greater than 1 in 10 to the 66 power can ever happen even if the universe lasted for an infinite number of years. Just the odds of our solar system forming accidentally by random chance from some laughable mythological explosion, our entire solar system with the sun and the planets and their moons etcetera etcetera, the odds of that forming accidentally by random chance are one in 10 to the trillionth power. End of The Big Bang Theory. It literally is a pile of horseshit. The theory that big bird built the Pyramids, by himself, actually has more credibility than the theory of the Big Bang. Amen and amen. Good luck and God bless.

-Can we finally put to bed this whole Nike /Colin Jackassernak Fiasco? Are you ready? Here's the final word. I just saw this posted on Facebook......

"Dear Nike: your claim that an anti-America, anti-police, pro-communist MULTI-MILLIONAIRE sacrificed 'everything' by disrespecting the country that made him rich is reprehensible. Sacrificed everything? Really? Go to Arlington National Cemetery. Speak to the widows and the children of our fallen military warriors. That is 'sacrificing everything!' Not some smug millionaire INGRATE who dishonors our veterans, service members, national anthem, and flag. Signed, We the People." AMEN!! Thank you, brother! …Colin Crapernack is just another stupid, left-wing indoctrinated, mentally poisoned useful idiot who, like millions and millions of other brainwashed suckers just like him, is stuck on Satan's black baby exterminating Democrat Plantation.

And how many idiots do we have to watch on our TVs running around like a bunch of jackasses with their hands up don't shoot? When that was nothing but a lie that had nothing to do with the reality of what happened there in Ferguson. Nothing whatsoever. But Satan and his lying demon rats own the media, the educational system and the entertainment industry so millions upon millions grow up brainwashed and stupid, ready to do the bidding of Satan and his child sacrificing Democrats. And that's what's going on in America Nike, you stupid ignorant uneducated tards. I would rather tape banana peels to the bottom of my feet, then wear anything made by Nike. They are piles of anti-American Communist China crap.

-Here's a joke. What do you call it when a moron, a whore, a Jew hater and a buffalo's ass-face walk into a bar? ..... The Squad. And what is the real name for the "Squad?" The four whores of the Apocalypse.

-Question. It is a known fact that in Washington DC roughly 90 to 95% of all the people there, of all Races, colors and nationalities are Democrats, or what somebody with a brain would say are baby killer voters. You know, like the Jew killer Nazi voters in Germany 80 years ago. Same difference, right? Right. …But that's not the question. The question is, and according to those percentages, is it safe to assume that the so-called Christian churches throughout Washington DC are filled with baby killer voters? And of course if you vote for baby killers, the demon rat Democrats, if you vote for murderers, then you of course are a baby killer and a murderer yourself. Right? Right. So are these churches in Washington DC Christian churches? Followers of

Jesus Christ? Or are they instead satanic churches. Followers of Satan, and his modern-day stealth, hidden, disguised, camouflaged religion of child sacrifice and lying and false accusations, child, sexual and marriage perversion, etc. etc. and other forms of pure evil that you would expect from one of Satan's religions. I'm curious. Whose churches are they? Jesus's? Or are they really Satan's churches. Let me know. I am seriously and sincerely searching for the answer to that question. And maybe you should be too. As well as all the baby-killing Democrap pastors and parishioners in D.C.....

-GOD IS LIGHT.. No, you don't understand. God actually IS light. And it can be proven..... Let's do an experiment. Out in space. (Hypothetically of course.) Way way out in space. A spot devoid of other objects where we would have the room needed. For in our hypothetical experiment we shall set up a hollow sphere 2 light years across. The material we use to construct this sphere will allow no outside light to penetrate it. Therefore it is absolutely "pitch dark" inside. In the center of our little sphere, floating there in the exact center, is a light source. A "bulb" if you will. About an inch in diameter. That shines in every direction. And then, on the inside of the sphere, covering every square inch of the inside of the sphere, let's place light sensors. Now, simple enough, very rapidly, for a split second, we'll turn the bulb on and off. And what happens? Well, about a year later, light hits the sphere, the inside of it, everywhere, every square inch. And every sensor would of course register this. That is the nature of light. But how is that possible?? Are you telling me that all of those "photons," the astronomical amount necessary to hit every square inch of the inside of a sphere roughly 2 light years across, came from that tiny little bulb in a split second?? How? Do the math. It's impossible. There's not enough space on that tiny little inch-across bulb for all of those trillions upon trillions upon quadrillions of photons. Hmmm… Is light then a wave? In space? Don't think so. No, light is obviously virtually infinite, from the moment of its inception. Therefore light has no "physical" property, no substance that we can define, in the material world. Light is God. And God is light. (And Life.) That is the only possible feasible explanation. Unless you can come up with a better one... (Lots of luck.) ... You are surrounded by light. You are surrounded by God. "God is light, and in Him there is no darkness at all." 1 John 1:5

-Here's how you "win" arguments with anyone on the left. You just say, "I don't vote for baby killers. I don't vote for baby mutilators. I don't vote for murderers. I don't vote for racist black baby exterminators." ....It's called planting a seed. You see they think they're very righteous but they have no righteousness or morality whatsoever because they are the filthiest thing on the face of the Earth. Even filthier than the Jew killing Nazis of Germany 80 years ago that everybody in this country spits upon their memory. They are baby killers. If you just repeat that over and over again it's planting a seed that eventually will burn inside of their brain. And there is no getting it out. ...."Well personally I don't vote for baby killers. I don't vote for murderers." …It doesn't matter what they say back… "Oh Trump this and Trump that." Just reply, "Well I don't vote for racist black baby exterminators." ...."But Trump this and Trump that and blah blah blah blah blah." ...."Well I don't vote for murderers. I don't vote for baby killers." Try it. You might like it. But don't ever say anything else to whatever response they have except for I don't vote for baby killers. Good luck and God bless. And for those who say that the baby killers and their voters are proud of it I would just like to respond that they never ever hear that they are baby killers. All they ever hear is that they are pro-choice. Or that they defend women's rights. Or that they are pro-abortion as if somebody is coming in too low for a carrier landing and is getting waived off. Abort! Abort! They never hear that they are filthy baby killers. And what I said above is the perfect way to plant the seed because instead of actually calling them the filthy baby killers and baby mutilator voters that they are, you put it in terms of what you yourself would never vote for. Which I think is a lot less offensive. But it plants the seed and that seed can never be removed. It will burn in there. And if you also couple it with the filth of the Jew killing Nazis in Germany 80 years ago which they of course hate, condemn and despise, then it really puts them in a mental conundrum. And keep in mind as well that more than just baby mutilators the Democrap Party is the party of bald-faced lying 24/7, of false accusations, of marriage perversion, child perversion, sexual perversion, America hatred and Israel hatred, demonic transgender mutilation of children, as well as the party of black baby extermination and pure racism and despising of the blacks. But because the Democrats own the dissemination of information in this culture, because they own 95% of the media and over 90% of the educational system and the entertainment industry, their voters are not aware of the pure evil and pure filth that they go out and vote for. But again, I concentrate on planting the seed of the baby killers that they are as Democrat voters. Hopefully from that seed they can then become aware of the rest of the pure filth that they vote for. Before they die and are in for a very rude awakening as they're thrown into the Lake of Fire forever and ever.

-As far as this bullshit that has been foisted upon us over the decades by the lying Democraps concerning the supposed *separation of church and state.*

It is just another of their many lies, and has nothing to do with the actual foundation of America, which they actually don't give 2 shits about except when they can distort it in order to tear down Christianity. (As the saying goes, the ACLU doesn't hate religion, just Christianity. Islam and Satanism are just fine with them in our public schools, buildings and institutions.) In the beginning the Bible was taught in public schools and prayer was offered at every single government function. The only thing that was supposed to be separated was the state *imposing* anything or any religion on the people. They were supposed to be free to choose their own form of worship. Christianity and the Bible were an integral part of our schools and our government in the beginning. Absolutely no separation there. That is a lie from Satan and his Demonrats in order to get God out of our schools and out of the Public Square. Of course the filthy child-sacrificing, baby-mutilating religions of Satan and his Democrats, and his clitoral mutilators, are welcomed into our "public" schools and our public square with open arms.

-Our public schools are NOT "government" schools. That is far too benign a term and therefore misleading as to what they really are. They are religious schools that push Beelzebub's religion. They are lying, left-wing Democrap indoctrination centers. Mostly what occurs there is "Vote Democrap" brainwashing. ("DEMOCRAPS GOOD! REPUBLICANS BAD! BABY MUTILATION GOOD! REAL CHRISTIANS WHO DON'T KILL BABIES, BAD! SEE SPOT RUN") They force the modern-day "progressive" (regressive) religion of Satan down the throats of our unsuspecting school children in our colleges, universities, elementary and high schools. And not-too-bright equally brainwashed servants go out and vote for it religiously every two years. Good grief.

-Coup d'état IMPEACHMENT HOAX." What the demon rats are doing in congress with their Trump impeachment fiasco is actually a textbook case that falls under the heading of domestic enemies. They are attempting a coup d'état, because they have no high crimes and misdemeanors. They don't even have any low crimes or misdemeanors. Therefore every one of those Democraps that signed on to this impeachment bullshit in Congress, and in the media, need to be arrested and tried for treason. Immediately. And then hung. On Primetime TV. From the trees lining the Mall in DC.

-Just a little truth bomb for our friends trapped on the Democrat/dark side of the aisle. Satan has total control of your mind. Better throw him out ASAP. Anyone who loves the lies of Satan that they were taught, kills his babies with their legislation and with their votes, and hates the truth of Jesus Christ their own Creator when it is presented to them, have given Satan total control of their mind. Our purpose here is to destroy the lies of Satan. Inside of your mind. Pay attention. You are a slave. Stuck on Satan's modern day, child-sacrificing, baby-mutilating, racist, black-baby-exterminating, child- sexual- and marriage- perverted, America- and Israel- hating, pure evil Democrap plantation. Our purpose here is to merely speak the truth and write the truth and if these posts and these writings cast you out of your safe places and throw you into a hissy conniption fit, with hatred and anger, and a spaz attack inside of your mind, then you can know that you are mentally poisoned with Satan's lies and your mind is under his control and you are his slave. I would suggest you get down on your knees and pray the most fervent prayer to Jesus that anyone has ever prayed in the history of the world, for him to come into your heart and mind and free you from the slavery to Beelzebub and his lies. "And you shall know the Truth (Jesus IS the Truth, and the Way, and the Life) and the Truth shall set you free." Otherwise a few seconds after you die you're going to sorely wish that you had. This is not an argument or debate between you and me or me and anyone else. This is just one fishing line after another that is thrown out there in order to catch people satanically and demonically possessed with lies. Pay attention. Your Eternal Destiny is at stake. Have a nice day. POWs in Hell never have a nice day. And that is what is at stake in this battle between God and Satan, between the truth and lies, between darkness and the light, between the lies of Satan that poison your mind and the truth of Jesus Christ your own Creator who you deny with your votes, and your allegiance to all the lies of Satan. (And PS, I am not a Republican. I am a follower of the truth, and a hater and exposer of the lies of Satan. There are many Republicans as well as Trump supporters who are not saved. Yet. Unfortunately.) Consider this a public service announcement. From God. ☒ Good luck and God bless.

-"I support my black friends, but not the BLM. I support my white friends, but not the KKK. I don't support hate groups." There's a new movie coming out about the morons of BLM. It's a loose remake on the movie The Color of Money. It's called the Color of Poop. The movie is in honor of the Democrats and their suckers (75% of whom are white) running around our cities beating people up and burning down buildings.

-The baby killers are following the Jew killers. They are one and the same. Hitler defunded the police so his brown shirt pig thugs could rape and kill the Jews and steal their businesses and their homes which is what the color of poop BLM is trying to do now, steal people's homes. The more things change the more they stay the same...

-The national slogan of the Democrats and their Ku Klux Klan used to be, "The only good nigger is a dead nigger." (Only now we know that the assholes were talking about themselves.) But they have changed their slogan in the last number of decades. It is now, "The only good nigger is a DemocRAT-indoctrinated, brainwashed, lie-filled nigger who will march out religiously every two years and do what they're told and vote for us." And keep in mind that they are brainwashed and filled with lies by a Democrat educational system, entertainment industry and media in order to keep them trapped on Satan's racist, black baby Exterminating, Democrat Plantation. They are voting for the slow genocide of their own race. 40% of all babies mutilated to death every year by the Democrat Party and their Planned Parenthood are black babies. Yet blacks are only 12% of the population. The founder of Planned Parenthood Margaret Sanger said, "We want to exterminate the blacks." Hillary Clinton said, "Margaret Sanger is my hero." And that's what 90 to 95 % of blacks go out and vote for, religiously, obediently, just like they've been taught to do, every two years. Somebody might want to ask LeBrainless James and the rest of those black and white and Hispanic Democratic voting geniuses to come back when they actually get a fing brain. Useful idiots.

-A few thousand years ago God said that Satan is the prince of the power of the air. But I would expand that verse just a little, for today, "he is the prince of the power of the airWAVES." Because a couple thousand years ago when that was written the readers did not understand TV, radio, the internet, etc. and the airwaves that produce that communication. But now we do, now we can see exactly what the Bible was talking about many years ago. And that it was written specifically for today. Beelzebub is the prince of the power of the airwaves. He controls the dissemination of information because 95% of the media are baby mutilating Demonrats. As well as over 90% of our educational system and entertainment industry, and 100% of social media before Musk freed twitter. And as such he controls the dissemination of information in our culture. And stupid people listen to him and go out and vote for him and his baby killers, and then they get thrown into the Lake of Fire after they die. OUCH!!

-The level of knuckle-dragging Neanderthal ignorance it takes to believe in some accidental random chance Earth Mommy that created all of life on earth in its astronomical and nearly infinite complexity is so idiotic that you could die laughing.

-As far as that disgusting picture of the Pope and a Muslim Iman smooching each other you have to understand that The Catholic Church never had anything whatsoever to do with Jesus Christ. The emperor Constantine just used the name of Christ to propagate their false pagan Roman religion and their perversion of the Gospel message. And no, I don't hate Catholics nor am I putting down individual Catholics that have been born and raised and trapped in that slightly false religion that is leading millions and millions and hundreds of millions to hell because they never accept Christ as their savior, they just trust in the Catholic church and their own good works. And also, BTW, I myself was born and raised Catholic so I know whereof I speak. There are many "good" Catholics, but a person's personal "goodness" and personal righteousness doesn't get them into heaven. ("All of your righteousness are as filthy rags." If you try to use that to get you into heaven rather than the righteousness of Christ. Which is righteousness indeed.) Only being covered in the righteousness of Jesus, only being washed in the blood of the sinless Lamb of God, a lamb without any imperfection (sin), will get you in. Does that mean that I am saying that there are no Catholics who are saved? Of course not, but that depends on each individual Catholics relationship to Jesus, and not on their church membership, or anyone's membership in any Christian church denomination for that matter.

-What is life? Well it's a story that the Creator wrote about six thousand years ago before the foundation of the world. Before the universe was created by Him and before he spoke all the living into being on this planet. He wrote all of this that we are reading now. That we are living inside of. And yes at the same time don't misunderstand, we all still have absolute free choice. Because if we didn't he wouldn't be a just God sending people to hell if they didn't have a choice. So we do have absolute free choice but at the same time he is absolutely sovereign over every minutest detail of every event the world over at every millionth of every second, because if he didn't have that kind of control then he wouldn't be God. So imagine if you will that you are in a three dimensional virtual reality that God made. You are inside of your brain that shows you what's going on in the world outside of you and lets you see and hear and taste touch and smell it also. All this really is just an image in the mind of God. ...And the really incredible thing is that I just wrote all of that without the use of drugs. HAH!!

-We live in a fallen world. There is never any real justice until after we die. Then believe me there is perfect Justice for everyone. Which is why Jesus died on the cross so we could escape what we all really deserve. And I am not making light of your suffering, or anybody else's suffering either. Some people suffer a whole hell of a lot more in this life than others do. But it's the suffering after we die that we really need to try to avoid. And that can only happen by faith in Jesus Christ. By being covered in his righteousness. By having on a wedding garment. (That's in Matthew in God's word.) Because this life even if you live for a hundred years it's just a split-second compared to Eternity. After you die whatever happened in this life, whether it's extreme suffering- the worst suffering, or great pleasure and success and happiness, really is of absolutely no consequence unless you have taken care of Eternity. So again I would just say, God answers prayer. Whenever anybody is willing to come to Him and ask. Good luck and God bless

-Was out hanging in California the other weekend, and saw that the Insane Demonrat piles of baby killing vomit just passed a new law allowing certain people, you know Democrats, the Clintons, demon rat donors, and the like, to have an exemption from their 2025 law that forced all Californians, except of course for illegals, into having sex with animals - their pets, barnyard animals, rats, other assorted rodents, zoo animals- you know, the law that Buttjudge, Camelsassface Harris, Corey IamJackassicus Booger, and other assorted Democrap, baby mutilating baboons passed in order to spread their Progressive love to all mankind and animal kind. Peta was a big proponent of this law. But now we see that they exempt their friends from it. Such Hypocrites they are. NAMBALA (The North American Man Boy Animal Love Association) is fighting this law in court now, as they think that no one should be exempt from having sex with children and animals, like they and their Democrat friends have been enjoying for years. Stay tuned to this station for further updates. If you don't puke to death beforehand. Hahaha. Stupid Democrat voters. (Yes. Time travel. ☺)

-Hey, did you guys hear that there was some kind of a debate on TV the other night. I heard it was called 10 jackasses. What was that all about? Were they talking horses or something? Or were they talking horse's asses? Yeah, that was it, 10 talking horses asses. ...What level of moron does it take to sit back and listen to 10 barking horses asses for hours?!? Oh, that's right, they're called Democrat voters.

-Anybody with a brain, which of course means anyone but the Demon-possessed-rats, and their voters, looks at all of the utter insanity that the Democrats are now for and just shake their heads!!?? "How can anybody be THAT stupid??" To want open borders which of course means no country whatsoever, but it does mean more gangs, more crime and violence and more stupid Democrap voters; the abolishment of ICE; the hatred of our protectors the police; sanctuary cities and states where they harbor dangerous criminals as long as they are here illegally; mutilating babies to death up to the moment of birth and even after; stealing our guns from us so we can no longer protect ourselves from the rapists and murderers whom they and their father Satan love; letting boys who say "I'm a girl" shower naked with the little girls in our publicly insane indoctrination schools, and letting them compete in, and therefore destroy, woman's sports; parading sick "transgendered" pedophiles in front of our 1st and 2nd graders; etc. etc. [And now, "defunding the police."] But you have to understand that they actually ARE insane. Because Jesus has given them over to what is called a reprobate mind which IS literally insane, and stupid, and absurd. And the reason Jesus/God does this is because they love the lies of their father the devil, and participate in his modern-day religion of child sacrifice. So God gives them what they want, a reprobate mind that is no longer capable of discerning truth from lies, common sense from utter stupidity, morality from evil. So they now call evil, good. And good, evil. ("Whoa unto them who do so.") If you understand that then you understand perfectly what is going on today. We live in an insane asylum, and the inmates have taken over.

-Defunding the police… "Honey, sweetheart, I'm sorry to say that who that is at the front door is 3 very large guys with guns and BLM and Antifa t-shirts that shouted if I don't let them in to gang rape you, and our 8 year old daughter (is she asleep upstairs in her bed?), then they will kick the door down. Now you know because we are good loyal liberal Democrats we are all for gun control, and our house is a gun free zone, and 911 just goes to voice mail now, so we have no means to protect ourselves. Soooo, I guess I'll just let them in, maybe they won't be as brutal then, and at least we won't have to fix the door. All right, here goes, and I sure hope you're not going to be too sore because we still have to go out and vote for Biden tomorrow…" ☺ Dumb Demonrat voting bastards.

-Now here's a question for Mr Buttjudge (Pete Buttigieg) It's a question that was never asked of our little presidential candidate, mainly because we have a Democrat media, which has no interest in holding any Demonrats feet to the fire. And what is that question? Well before I ask it, let's fill everyone in… In case you haven't heard, there are those infamous poo-poo parties, in San Franshitco, and elsewhere, where homosexuals will get together at a friend's location and get naked down on all fours and crawl around in a circle, face to one another's ass (like circus Ellyfonts) (see Lord of The Rings), and poop into one another's mouths. …So as an independent journalist my question is, does Buttjudge eat the shit out of his husband's ass? Inquiring minds would like to know. Let me know what his answer is. And nothing against our homosexual brothers, most of whom I would venture to guess, do NOT engage in the San Franshitco poo-poo party. But does Buttjudge and his husband?? Ask him. And then, stupid, ignorant, deceived, uneducated Democrats who voted for him, are you going to teach your sons to do the same????

- The C word… Wouldn't it be perfectly acceptable if someone on national TV said this, "Michelle Obama is a cunt!" (Assuming she IS a Michelle down there, and not a Michael.) Or, "Chelsea Clinton is a cunt!" Or, that fat, disgusting, racist, baby-butchering pig who stinks up Georgia, "Stacey Abrams is a cunt!" Would that be OK?? Of course it would! Because that's what Bill Maher called Sarah Palin and that was completely acceptable. None of the baby-killing, lying pigs in our national media even batted an eye. Sarah, you see, doesn't kill her babies. (Completely intolerable to the demonically-possessed Democraps.) Sooo, have at it… "Nancy Pileofshitsi is a cunt!" "Bill Maher's wife is a cunt!" "Mrs Biden is a cunt!" Is this fun, or what?? Thanks Bill.

-The Elderly The Democraps care so much about the elderly… And they show their great care and concern by trying to kill as many of them off before they are born. That saves them a lot of money down the road…

In case anyone is wondering why the Jews have been massively persecuted everywhere they have been exiled over the last 2000 years, including what happened on Oct. 7th, 2023, just read Matthew 27, verses 24 and 25… "then Pilate washed his hands before the multitude saying, 'I am innocent of the blood of this just person.' And all the Jews answered and said, 'His blood be on us and on our children.'" OOPS!! OUCH!! To say that was the BIGGEST mistake in the history of the world would be a vast understatement…

-The Democrats are obviously pure evil, worshippers of Satan, baby killers, racist against the black race, black baby exterminators, child perverts, marriage perverts, sexual perverts, transgender-child-mutilating absolute spastics, America haters, etc. etc. But the problem is the bamboozled uneducated brainwashed baby killing suckers that go out and vote for them and seal their eternal fate in the Lake of Fire by doing so. Maybe some political party should try and wake them up. If they have any balls whatsoever

-There has been a few times in the past where either a couple of the evolutionists or a few of the Trump hating Democrats have tried to criticize my posts, not by their content, but by accusing me of having "anger issues," and that I need help etc. etc. Well in response to these geniuses let me just say for the record that I do have anger, and more than that I have hatred. I have anger and hatred for the lies of Satan that cover this earth like a blanket, and that these critics still believe. The lies that we all were taught but many of us thank God have escaped from. The evolutionary lies, the Big Bang lies, the "hey let's all go out and vote for the baby mutilating, lying, false accusing, Trump deranged, satanic Demonrats" lie because that's what we were told all of our life to believe and do. Now I will admit that sometimes that anger and hatred for the lies of Satan does spill over into anger for his stupid followers that still propagate those lies. And that is not right. We are supposed to hate the sin but not the sinners. And for the most part I, and my posts, do love the sinners. Many times I love them a little too much by smacking them upside their brainwashed heads. Hahaha. But that's tough love. But here's my peace offering to those critics... you pray for my anger issues, and I'll pray for your satanic indoctrination. Deal?

-All I can say is keep killing Jesus's precious babies you geniuses. Keep voting to mutilate them to death. And watch what happens next...

...And for God's sake, it's not "fake news." It's lying Democrat news. Lying Democrat news. Repeat after me. Lying. Democrat news. And somebody tell Trump.

-Atheism is for idiots. With all due respect of course. It's no different than if you talk to somebody sitting in their car and they said that there is no such thing as Detroit or Japan or any other manufacturer and that the car just made itself. You would know that they were kinda dumb. And as far as the origins of the universe and life on Earth it is a scientific fact according to all the evidence and according to the very laws of science that both the theory of evolution and the theory of the Big Bang are total piles of scientifically impossible horseshit. The universe didn't make itself and that's a fact. Life on Earth didn't make itself and that's another fact of science. So all this had to be made. Just like a car has to be made. Or even a stupid paperclip has to be made. So again, and with all due respect, atheism is for idiots.

"Laughing" by David Crosby (Slight return)

"I thought I met a man. Who said he knew a man. Who knew what was going on.

I was not mistaken It was Jesus That he knew.

And I thought that I'd found a light To guide me through my night and all this darkness.

I was not mistaken. It was the Light of Christ. That I saw.

And then I thought I'd seen someone, who seemed at last, to know the truth.

I was not mistaken. Only a child of God laughing. In the joy of the Son."

-The Democrats asked their father Satan to teach them how to pray, and this is what he said.... "Pray to me in this manner... Our father Beelzebub, who art in hell, revolting be thy name. Thy baby killing Kingdom Come, thy lies and false accusations by your Demon-possessed-rats and their Muslim brothers be done, on Earth here as it is in hell. Give us today the ability to destroy all that is good and honest and pure and holy, and let us falsely accuse and disparage all of those decent people who try to stop us, as we brainwash their children into becoming filthy evil satanic lying BLDM baby killers like us. And let us continue to temp the fools into voting for and supporting us and giving us the power to accomplish all of your evil designs for this once great country. In Satan's name, AMEN."

-Imagine if you will a huge mountain of stinky wet shit and diarrhea, and it oozes out from its native Middle East lands and slowly spreads across the world in order to stink up decent countries and decent people with its noxious, odorous, women subjugating, clitoral-mutilating ideology/religion. Imagine that if you will. And you get a good grasp of what's going on, and is aided by the Demonrats in this country.

-"Defund the police" is just another way of women saying, "Hey!! Please come rape my ass."

BLM... Burn. Loot. Murder

BIDEN...

B. Biggest

I. Idiot

D. Democraps

E. Ever

N. Nominated

-Time for blacks to get off of Satan's Democrat Plantation. Time for whites to do the same. Time for stupid Trump haters like the Philadelphia Eagles and LaBrainless James and Seth Cur and Popobitch and Steve Cur to get some actual political edgumakation and get off of Satan's racist, black baby exterminating Democrat Plantation. Which is purposely and premeditatedly perpetrating the slow genocide of the black race. Because Margaret Sanger the founder of Planned Parenthood years ago said "we want to exterminate the blacks," and Hillary Clinton just recently said "Margaret Sanger is my hero," and 40% of all babies mutilated to death by the demon rats and their Planned Parenthood are black babies, yet blacks are only 12% of the population. Do you need a calculator LeBron? You BLM dumbass. And that's not racist, stupid. That's called the truth. And the truth isn't racist. It's the truth.

-Like David, Trump is a man after God's Own Heart. Which is why Satan and his baby mutilating demonic Democrat followers hate him with such a satanic passion. People are actually desperate for the truth in this country, and of course in this world as well. But in this country they've been lied to all of their life because Satan and his baby killing Democrats own over 90% of the media, the educational system, and the entertainment industry. This country is totally fucked if we don't get the truth in front of these people.

-Isn't it hilarious, and absolutely sick, watching the brainwashed Democrat ignors tearing down statues and tearing down history like piles of stinky, self-righteous filth. They are so much better than those slaveholders from hundreds of years ago. They just are slaves of the modern-day Democrat slave holders, useful idiots imprisoned on their father Satan's modern day Democrap plantation, where black babies are mutilated to death at four times the rate of white babies. These vile, black baby exterminating piles of brainwashed, history destroying Democrat dogshit have no morality or righteousness whatsoever. They, and the ones who brainwashed them, their Democrat professors and teachers, are a stench upon Humanity. But thank God they will all be gone very shortly, after they have eventually completely destroyed America and usher in the seven-year reign of Satan and his Antichrist. Their reward for that accomplishment will be a plunge into the Lake of Fire, molten lava: Sulphur boiling and burning at 1000 degrees, forever and ever. No statues to tear down there Satan's suckers. Just your screams and excruciating pain and absolute horror, forever and ever. ... Unless you repent...

-How can a Jew even be a Democrat for God's sake? They are filthy, disgusting, Jew-haters and Israel haters. In addition to perpetrating a modern-day Holocaust that has mutilated 60 million babies to death. After the Holocaust that occurred in Germany 80 years ago you would've thought they'd be a little sensitive to being a part of another sickening Holocaust. You'd have thought wrong. And don't try to accuse me of being anti-Semitic. There is no stronger supporter of Israel and the Jews than myself. This is the truth. And the truth is not "anti-Semitic," or anti anything, except for being anti lies of Satan. And it is those lies that keep people trapped on his modern day, disgusting, filthy, baby mutilating, America hating and Israel hating Democrat Plantation. Wake up! Jews and Gentiles. Blacks and whites. Hispanics. Get off! ...Before you die...

-Masks are for idiots. Masks are harmful. Masks don't work. They just block oxygen. They can't block a virus. We've all been had. SUCKERS!!! All businesses are essential. Democrats are non-essential. Democrats are the virus. Vaccines are the cause of autism. Bill Gates dad was a monster. Bill Gates is a monster. The Nazis were Jew killers. The Democrats are baby killers. What's the difference?

-THE DEMOCRATS ARE THE RACISTS!! NOT THE POLICE. WAKE UP, SUCKERS!!

The Democrat Party and their black baby exterminating Planned Parenthood are now and always have been the party of blatant racism and complete contempt for the blacks.

-The hoax of man-made global warming is for the hysterical true believers among us (who just believe whatever idiotic lies they were taught.) (You know, like the one where you get 72 virgins by flying highjacked planes into tall buildings.)

-"Blacks Lives Matter" doesn't give a rat's ass about black lives, or they would have gone to Chicago a long time ago. And Baltimore. And they would march in front of the black-baby-extermination clinics of Margaret Sanger ("We want to exterminate the blacks") and Hillary Killery ("Margaret Sanger is my hero.")

-What do all these people have in common? Quid pro quo Joe. Nancy Pileofshitsi. Adam Schiff for brains. GoNadler. Upchuck Schumer. George Sorass. Maxinsane Waters. LeBrainless James, Liarwatha. Commie Sanders. Joyless Bitcher. Bigtoe O'Rourke. Buttjudge. Whoopee Cushion. Don Liarman. Donnie Bitch. Sewer rat Sharpton. IamJackassicus Booker. Camelsassface Harris? They are all domestic enemies. They hate America just like they were brainwashed to, and they work tirelessly for its collapse. And didn't I read somewhere that we are supposed to defend this country against ALL enemies, foreign and DOMESTIC??? Hmmm…

-Someone suggested that our society has lost its compass. No. I respectfully disagree. Our society is run by baby killing, lying, false accusing, Satan worshipping Democrats. They own 95% of the media and over 90% of the educational system and the entertainment industry. Our society hasn't lost its Compass. Our society has been overrun by piles of pigshit.

-Here's some wisdom from one of my friends on Fascistbook, Essell Cee....

"The problem is, not enough people are versed on the diabolical nature of the Democrats; their racist history; their exploitation of minorities through their identity politics scheme (NOT a philosophy); their disloyalties to their constituencies, loyalties sold to the highest bidders whose purpose is not conducive to the betterment of America. In short, many Americans are ignorant to the bone, done through their slavish dedication to the very source of "information" that conflicts with a well-informed voter and that would allow them to make intelligent decisions on who best represents the interest of the citizenry! We all that are concerned about the dumbed down status of our fellow citizens, have to continue to fight exhaustively against the ruinous propaganda spewed out on a daily basis by the media enemies of America! Keep it up, Danny." God bless you, Essell.

-Question. What exactly is the difference between being a member and or a supporter of Black Lives Don't Matter (it's accurate name) and the

Nazi/Fascists of "Antifa," and being a member and or a supporter 80 years ago in Germany of the Brown Shirts and their Hitler Youth? You're right. None whatsoever.

-Elizabeth Warren is no longer claiming to be an American Indian. She now says that she is a bona fide black Samoan rhesus monkey. (Don't ask me.)

-The demon-possessed rats are baby killers. That says it all. That's all you need to know. Would you spend any time debating the Nazi Jew killers in Germany 80 years ago? Of course not. You're not that stupid. Well somebody explain to me what's the difference between the Nazi Jew killers in Germany 80 years ago who slaughtered six million Jewish men women and little children, and the Nazi demon-possessed Democrats in America that have mutilated 60 million babies to death, and are still going strong, and are so proud of it? There is none.

-The Democrats running for the presidency, along with their demon rat friends in Congress in the media in our educational system and in Hollywood are brainwashed tards whose minds are filled with lies and demonic hatred for all that is decent and honest and right in this country. They literally are pile of refuse. Now let's all go out and vote Democrat.

-"And the Sea gave up the dead which were in it, and death and Hell delivered up the dead which were in them, and they were judged every man according to their works. And death and hell were cast into the Lake of Fire. This is the second death. **And whosoever was not found written in the Book of Life was cast into the Lake of Fire**. .... And he said unto me, it is done. I am Alpha and Omega, the beginning and the end. I will give unto him who is athirst of the Fountain of the Water of Life freely. He who overcomes shall inherit all things, and I will be his God, and he shall be my son. But the fearful and unbelieving and the Abominable and murderers (i.e. the baby mutilating Demonrats and their voters) and whoremongers (which includes the sexual and marriage perverted Democraps) and Sorcerers and idolaters and all liars (which is the Democrats and their media, educational system and entertainment industry) shall have their part in the lake which burns with fire and boiling sulfur, which is the second death." ...Somebody might want to tell the Democrats. And their voters. Before it's too late.

-The Trump haters are just fools deceived by their father Satan, who does everything to keep them trapped on his baby mutilating, child sacrificing, bald-faced lying, false accusing, child perverted, sexually perverted, marriage perverted Democrat Plantation. It is sucking them down to an eternity in Hell. Because there will be no murderers in heaven, and if you vote for the murderers guess what you yourself are? Yet they think they are so smart and so righteous. But they are ignorant baby killer voters and have no morality whatsoever. ... Hey! Here's a thought! Maybe somebody oughta try to wake them up! Maybe the Republicans and conservative commentators in the media can start calling the Demonrats and their voters what they are: BABY KILLERS!! ... Or we could just let them go to hell.

-There's nothing more disgusting and evil than somebody who's being falsely accused, like Kavanaugh by the satanic Democrats. And like Trump's been falsely accused for years by the DemonRATS.. Trump's a racist. Lie! Republicans are racist. Lie!! Trump's a liar. Lie! What a HOOT!! He's being falsely accused of lying by some of the biggest liars on the face of the Earth. That's all they do is lie 24/7. But to those assholes, Trump is the liar. How hilarious. And Russia collusion. Lie!. Trump needs our prayers cuz he's being attacked by Satan and all of his demon rats and being falsely accused constantly. "The Demonrats lies they are drawn, their curse it is cast. The God haters now will later be last, as the perverts now will later be past. The Lord is soon rapidly returning. And Satan's baby killers now will later be damned. For The World is Soon-to-be A-Burnin... ...Please get out of the way if you can't free your mind. For the Lord, He is Returning." Dylan. Slight return

-"I got my own life to live. I'm the one who's got to die when it's time for me to die. So let me live my life the way *the Lord wants me* to." Hendrix. Slight return

-And on another note, what can be said about the brain-dead imbecile Colbert and the trained left-wing-indoctrinated seals in his audience? ...Oh, wait a minute. I just did.

It is pathetic and reprehensible that Republicans in Congress and conservative commentators have not screamed from the rooftops that the treasonous Democrats have kept roughly 50,000 veterans homeless, while giving food, clothing, shelter and healthcare free to millions of illegal immigrants (AKA Democrap votes). (Stupid, uneducated people vote Democrap.)

"BE STILL. And know that I am God. I will be exalted among the heathens. I will be exalted in all the Earth."

"Turn off your mind, relax, and float downstream. This is not dying. This is not dying. Lay down all thoughts surrender to the *Lord*. This is being. This is being."

"Row row row your boat gently down the stream. Merrily merrily merrily merrily life is but a dream."

"Bring into captivity every thought to the obedience of Christ" 2nd Corinthians 10:5

"Words. The way is beyond language. For in it there is no yesterday, no tomorrow, no today..."

"Gone. Gone. Gone beyond. Gone beyond beyond. All hail the goer."

"This is what the Sovereign Lord, the Holy One of Israel says, 'In repentance and rest is your salvation. In quietness and trust is your strength. But you would have none of it.'" Isaiah 30:15

Hmmmm…

-"Black Lives Matter" (BLM: BoweL Movement) is the stupidest possible title for any organization in the history of man. A bunch of people sauntering down the street chanting "black lives matter" is as stupid and ignorant as a bunch of people sauntering down those same streets chanting, "The Earth is round." It's like, hello you jackasses, everybody already knows that. Everybody with a brain already knows that black lives matter, as well as police lives, and all lives. Everybody, that is, except you and all of your stupid Democrat, racist, black baby exterminating friends. It's almost as hilarious as the title "Antifa," which stands for a troop of mentally deranged, left-wing indoctrinated morons who are the real Nazi fascists. The other really hilarious thing about the geniuses of black lives matter, whether they're black, white, Asian, Eskimo or Martian, is that they are all Democrats. And yet they're too ignorant and uneducated to know that it is the Democrat Party that is now and always has been the party of blatant racism, and total mocking distain and contempt, for blacks. They are the party of the Civil War, and the KKK afterwards, and Jim Crow laws, and now the slow genocide of the black race by their orchestrated ongoing Planned Parenthood campaign of black baby extermination. Roughly 40% of all babies mutilated to death every year are black babies yet blacks are only 12% of the population. Hmmmm. Maybe LeBrainless James needs a calculator. The rioting, looting and killing useful bozos of BLM, along with their lying Democrat media, are using the death of Floyd (who was NOT killed by the police, but died of drug overdose) to scream that the police are racist and are perpetrating genocide against the blacks. Which anyone with an actual brain knows is a totally absurd lie. All the facts and statistics confirm this. But the lying DemonRATS are the real racists and are perpetrating a real genocide with their Planned Parenthood situated predominantly in poor black neighborhoods, and it's all being supported by the morons of BLM and their media. The term "useful idiots" comes to mind. And now satanic, black baby exterminating, Democrat-supporting corporations are giving millions to these BLM racist, mentally disturbed, LaBrainless James morons that are burning down their own inner cities across America. We live in an insane asylum and the Democraps want to take away your guns and now are calling for the defunding of all police, while at the same time getting rid of prisons. My only question is, how many uneducated fools are still going to march out and vote Demonrat?? They need to be put in mental institutions and not in office!!!

-Better arrest all the traitors that perpetrated a coup d'état and hang them all from the trees lining the national Mall in front of the capitol. For all the country and all the world to see. Better arrest all the pedophiles and child torturers, baby eaters like Hillary and her friend Podesta and Huma Aberdeen, and adrenochrome blood drinkers that run Hollywood, the Demonrat party and the media in America. Better shut down the Nazi sensors that run Facebook and Twitter and YouTube and Google and replace them with Americans that don't have a satanic, Demonrat agenda. Better shut down the national lying Democrat media because it has nothing to do with freedom of the press. It has become the propaganda arm of the America-hating Democrat Party. Better get rid of all the disgusting lying, false accusing, baby mutilating, Satan-worshiping, Socialist Communist pigs that run our educational system from elementary through high school and college and university that have been poisoning the minds of our young people for decades. Now if all these things are not done, and not done with extreme prejudice, you can not only kiss this country goodbye but you can kiss yourselves goodbye. Because they will continue to let the criminals out and the criminal illegal aliens in and then they will take away your guns with police showing up at your door. So either you die or you give up your guns and you let criminals and criminal illegal aliens break into your house with guns (they will still have them, of course) and rape your wife and your daughters in front of you while they torture you all to death. That's your choice conservative America. Pray for Gideon. (See the Book of Judges. Chapters 6, 7 and 8.)

-I never did grieve for Floyd. Cuz I'm not a jackass. He was nothing but a career criminal. Good riddance. The moment I saw it I grieved for the police cuz I knew what the racist Democrap piles of pig vomit we're going to do. And of course they did. With their ownership of the media. Floyd died of an overdose of fentanyl. That's why he "couldn't breathe." A typical symptom of that overdose. This was confirmed by the autopsy. That he had massive levels of that drug in his system. As well as other drugs. Police, and law enforcement, and military police, and even bounty hunters all over this country and all over the world use that technique of kneeling on the neck of an unruly, uncooperative subject. It doesn't block the trachea and doesn't cut off breath. But that's the truth which the filthy, America destroying DemonRATS and their media have no interest in reporting. Because then their moron followers wouldn't BLM, Burn, Loot and Murder (BoweL Movement) and they couldn't assault the police. Morons. Fools. Suckers. Useful idiots.

October 7th, 2023. Watch the ½ hour video of Satan-controlled demons running wild, preparing themselves for an eternity in the Lake of Fire, and then tell me they were not born into a filthy, clitoral-mutilating religion of the demonic entity, Allah, and the prophet of hell, Mohammed, and raised to hate all Jews with a mindless spastic uncontrolled hatred. Tell the 1,400 slaughtered Jewish men, women and little children, who had their babies heads cut off and watched them burned alive along with their mothers and fathers. What inhuman demon sticks babies in ovens?? Tell that man who had his head hacked off with a garden hoe. And the woman who had her head chopped off with a shovel. And the women who had their genitals cut off by demon pigs of the pig Mohammed. Tell those little girls and women who were brutally gang raped and then taken to Gaza and marched down the streets bleeding profusely from between their legs, with those fucking mindless hysterical "innocent Gaza civilians" cheering them along. (And all because of the liberal Jews and their "gun control" so nobody was able to defend themselves and their loved ones…) And did you hear about how the hell-bound Hamas pigs showed that ½ hour movie one night to the cheers and applause of multitudes of "innocent Gaza civilians." (Personally I would have dropped a Daisy Cutter to aid in the festivities.) Tell me all about that "religion of peace" you fucking Democrat morons. The absolute horror of being born and then indoctrinated into a demonic Jew-hating religion from hell is obviously terrible beyond belief. And the Democrat mental slime has allowed these people to flood into our country by the plane loads. Is it too late to get them out? And now we have allowed them and their lies about the Palestinians in Gaza to poison the minds on our college campuses and we have to watch these retards protest constantly in support of the horror of October 7th!! Come quickly Lord Jesus…

-The Taliban has come to America. A cockroach infestation in our cities and College campuses. Protesting for their Hamas heroes disguised as support for "innocent Gaza civilians". (BTW did you know that "Hamas" roughly translates as "Asscrack"?) It seems tearing down statues and destroying our countries history wasn't enough to satisfy them. So they have come back for more. I think what they are really protesting is the fact that they can't laid. Not by any decent woman. Only left-wing indoctrinated whores.

-This is from Tucker Carlson..... "We have spent an awful lot of time over the past few weeks trying to figure out what is happening to our country. At this point it's pretty clear that nothing is what we are told it is. These are not protests. This is not about George Floyd. This is not about systemic racism whatever that is. America is not a racist country. You're not a bad person for living here. These are definitely not protesters. They are not even rioters. They are the armed militia of the Democrat Party. They are working to overthrow our system of government. They're trying to put themselves in power. That's all obvious now. It's genuinely sinister. But it's also very amusing. These people are idiots. For real. The angry children you watch set fire to Wendy's and topple statues and scream at you on television day after day are truly and utterly stupid. There probably has not been a dumber group gather together in one place in all of American history. They know nothing. They couldn't tell you who George Washington was. They don't know when the Civil War was fought. Probably not even to the century. They say they oppose racism and then they rip down monuments to abolitionists. A lot of the very stupidest people are "well-educated." They have all the worthless credentials. They went to Duke. They work at a nonprofit. They have a high-paying job as a digital marketer for Nike. They are not impressive at all. They are incredibly dumb. They fall for any lie no matter how preposterous it is. The more preposterous the lie the more likely they are to fall for it..."

-The Muslim religion is a stinking disgusting clitoral mutilating, women subjugating, insane, genocidal, worshipper of Satan and the prophet of hell Muhammad and his demonic entity Lalah. But besides all of that, they are really great. They actually cut the clitoris off of hundreds of millions of little 8 year old girls, and throw homosexuals and lesbians off of buildings. On their head. And then go down and rape the ones who are still alive, and writhing in excruciating pain. Someone might want to tell Don Liarmon. They are far beyond repulsive disgusting and revolting. Thanks to the demonrats they have infested our public schools. And watch out little girls! They are coming for a clitoris near you.

-Wrong Donald. It's not fake news. It's lying Democrat news. Lying Democrat news. Lying Democrat news. Lying Democrat news. Lying Democrat news. What is it again? Lying Democrat news, lying Democrat news, lying Democrat news. One more time? Lying Democrat news, lying Democrat news, lying Democrat news. Get it?

-Have you heard the latest announcement? The Democratic presidential candidates at the next debate are going to stand up together in solidarity and say that if any one of them gets elected that person is going to pass a new progressive, woman's health law making it okay for them to kill their babies up to 5 years old. ...As well as giving free healthcare to New Zealand.

-Hey! I've got it! Here's a really great idea... Let's not help the cops find and arrest any of the rapists, murderers, house invaders and other piles of pigshit in our community. In fact, let's make sure nobody does help the police by calling anybody who does a "snitch." And let's mock them and shame them and if we find out who they are let's beat them up and even kill them. Isn't that brilliant? And then we can watch our homes get invaded some more and our moms get raped and our sisters get raped and our daughters get raped. Aren't we smart? ...Fing retards.

-OH WHERE OH WHERE HAS MY CLITORUS GONE.

OH WHERE OH WHERE HAS SHE GONE.

Maybe we should ask Mohammered. He and his snips are having a drink down at his local watering hole, More Hammered.

OH WHERE OH WHERE HAS HER CLITORUS GONE?

OH WHERE OH WHERE DID IT GO?

Mohammered had a problem with the female orgasm, so their little girl's CLITORUS'S HAD TO GO.......

The religion of Piece. Yea, RIGHT. What a crock. Piece of genital flesh. How revolting and satanic.

-PRESIDENT DONALD TRUMP!!!, FOR JESUS GODS SAKE, I've been saying for well over a year now to please stop calling it fake news. That's not what it is. It's LYING DEMOCRAT NEWS. That's what it is for God's sake. Please call it what it is. Lying Democrat news. Filthy Democrat Liars that own our news media. Lying Democrat news

-"I, Bill Clinton, did not rape that woman, Juanita Broderick, in a hotel room in Arkansas. Yeah RIGHT! Ha-ha. And I got away with it because we baby mutilating, racist demon rat Democrats own the media. Ha-ha again. #Metoo?!?!@$#$ Ha-ha. What a crock of shit. What a joke. And you dumb asses fell for it. #Metoo this, bitches. What a joke we've got on all of you. Like we give a shit about women who get sexually assaulted or raped. Ha-ha. If we did do you think I'd be running around free as a bird giving speeches to all the hashtag me-too shyte-for-brain demon rat Democrats? Of course not. The jokes on you America. SUCKERS!"

-The squad has changed their name. They now prefer to be known as the four whores of the Apocalypse... however, in light of recent developments, they are contemplating changing their name again, to the baby eaters. Stay tuned to this station for further developments.

-"The rich man died and was buried. And in hell he lifted up his eyes in torment and asked Father Abraham to have pity on him, and send Lazarus to dip the tip of his finger in water and cool his tongue, because he was in agony in the fire." Somebody might want to warn our Democrat voters about where they are heading for all eternity. Because there will be no murderers in heaven. And if you're brainwashed enough to go out and vote for the murderers, the baby mutilators, the child sacrificers of old, the Satan worshippers, the demon-possessed rats, the Democraps, then you yourself are a murderer, and you have consigned yourself to your Eternal fate, unless you repent and come from Darkness to the light, from the lies of Satan to the truth of Jesus Christ. And the really hilarious thing, but it's not that funny, is that 4 million of these voters actually think they are Christians and followers of Christ as they go to church on Sunday and pray and worship and give tithes and offerings, and jump up and down shouting "Jesus, Jesus.". But they are in for a rude awakening when they die. "It would be better for you if a millstone were tied around your neck and you were cast into the sea, then you should mutilate millions of my precious babies to death with your votes". "Depart from me you workers of iniquity, I never knew you."

-"MAN-MADE GLOBAL WARMING" HORSESHIT SCAM

that's what it is. Somebody might want to inform little Thunberg. They are hysterical, brainwashed little useful idiots of the Democrats and their father Satan and their attempt is to destroy America and capitalism, so they can remake the world in the image of Venezuela, and North Korea, and China. They don't give a rat's ass about the planet, or the environment. (Or they would rail against China, and not Trump.)

-Here's a funny story. It's absolutely true, but totally pathetic. 4 million so-called Christians march out religiously every two years and vote for Satan and his modern-day, stealth, hidden, disguised, and camouflaged religion of baby mutilation, child sacrifice, bald-faced lying 24/7, false accusations (Kavanaugh, Trump), child sexual and marriage perversion (Epstein and friends, Weinstein, Hollywood), America hatred and Israeli hatred etc. etc. Obviously by voting for and supporting this glaring sinful activity, and being totally smitten with the filthy pure evil lying disgusting false accusing leaders of this Democrat satanic religion, they show themselves to really have nothing whatsoever to do with Jesus Christ or his church here on this Earth. Of course they are deceived. Heavily. And they are part of the religion of Satan. Also, and while we're on the subject, Blacks go out to the tune of 90 to 95% religiously every two years and vote for the racist black baby Exterminating demon rat Democrats. Even though the Democratic Party is now and always has been the political party of total racism, hatred and laughing out loud at the blacks. Somebody might want to tell LaBrainless James and his sidekick Seth Cur. But since Satan and his demon rats (Satan being the prince of the power of the airwaves) own 90 to 95% of the national media, the educational system, and the entertainment industry, Beelzebub and the Demonrats control the dissemination of information in our culture, and people only know what they're taught. ..."Satan's lies cover this Earth like a blanket and people believe whatever they are taught." Somebody might want to tell clueless La brainless James, Don Liarman, Maxinsane Waters and all the rest of the poor fools stuck on Satan's racist baby killing, black baby exterminating Democrat Plantation. "It would be better for them if a millstone were tied around their necks and they were cast into the sea than they should mutilate millions of my precious babies to death with their votes." And if those four million so-called Christians don't repent and change and get off of Satan's plantation, then these are the words they're going to hear right after they die,. "Depart from me into the Lake of Fire for all eternity, you workers of iniquity, you voters of pure filth, I never knew you". This has been another public service announcement. From God. ...You're welcome

- Where did Common Sense go? It does not exist in the brains of the baby killing Democrats and their voters. Because God has given them over to what is called a reprobate mind. It is a mind that is literally bereft of any common sense whatsoever and one that is totally insane along the lines of the important subjects where they have chosen Satan and his lies over Jesus and his truth. Angels protect and defend babies. Demons mutilate them to death.

-"Many will say to me in that day, Lord, Lord, have we not prophesied in thy name? And in thy name have cast out devils? And in thy name done many wonderful works? And in thy name gone to church regularly? And in thy name sang hymns, and tithed, and taught Sunday school, etc. etc. And then will I profess unto them, I never knew you: depart from me, you that work iniquity." Matthew 7:22-23 KJV/Dan's Amplified Version

"It would be better for you if a millstone were tied around your neck and you were cast into the sea then you should offend millions of my precious little babies by mutilating them to death with your votes." "Depart from me you worker of iniquity, I never knew you."

-As far as who or what created the universe and life on Earth, you have to understand that the Creator, God, call him or her or it whatever you want, is not limited by time or space, nor does He exist within the confines of time or space. He exists outside of those confines and is free of them. Time and space are merely creations that came into existence when the universe was formed about 6,000 years ago. Truth and love however existed before the universe was created because they are a part of His character, of His being. Trying to understand where all this came from and the nature of the Being that created it all, by looking within the confines of the universe, within the confines of time and space, is like trying to find the manufacturer of your vehicle by looking inside of it. And not looking outside of it, to Detroit or Japan or what have you. This error is the major flaw in the illogic of materialists, evolutionists and atheists.

-You must wear a mask. Unless you are rioting, looting and killing then it's okay not to. You must social distance. Unless you are rioting, looting and killing then you can get as close together as you want. You must not gather in large crowds of any more than just 10 people or less. Unless you are rioting, looting and killing then it's okay to be packed tightly together in crowds of thousands and more. Idiots. We've been had. We've been played. It had nothing to do with our "health." It had everything to do with destroying this country economically and putting as many decent people as possible out of work so the Demonrats could get their dementia senile retard and his hosebag sidekick elected to further destroy America.

-Life

-Well the Washington Redskins have finally succumbed to the pressure from the mentally deranged, America-despising, brainwashed DemocRATS and are changing their name. It's a toss-up between The Washington Kneelers, or The D.C Chinese Communists. Go online and cast your vote. (BTW, idiots, for the record it was the AMERICAN INDIANS who called THEMSELVES Redskins. It was never a slur. And the majority of American Indians today could care less about that football team being named after them. Idiots.)

-The baby mutilating Democrats and the clitoral mutilating Muslims are one and the same. The both have the same father, Satan.

-We have monsters living among us. Democrat monsters and Moose Lamb [I used this many times instead of "Muslim" in my Fascistbook posts so it wouldn't get flagged and the pigs wouldn't ban me] ...monsters. And they are here to destroy America. Because that is the desire of their father Satan. He needs to destroy America so he can destroy Israel before the Lord returns and therefore he can block prophecy from occurring and he can keep himself from being thrown into the Lake of Fire forever. That is what is going on in a nutshell. They are baby killing monsters, demon possessed, and they own the media, and our educational system, and our entertainment industry, and of course the demon rat Democrat Party. They are monsters just like the demon-possessed Nazi Jew killing monsters from 80 years ago in Germany. And really stupid uneducated people go out and vote for them, and give them power. Power to destroy America and to do the will of their father Satan. How sad.

-There is a horrible environmental catastrophe going on in California and it's caused by all those homeless in Los Angeles and San Francisco shiting and pissing all over the streets which gets washed directly into the bay and the ocean, as opposed to going into sewage treatment plants which it does when people shit and piss into their toilets. It has been documented that marine animals are all of a sudden dying at an alarming rate. But it's funny how you don't hear any of this from the "environmentally conscious" Democrat media. They don't give a rat's ass about the environment. They just want to get rid of capitalism.

-Check out the Book of Revelation and see what it says about the plagues that are coming in the end times. Half the world's population will perish due to famine, plagues, war, pestilence, and starvation etc. etc. This coronavirus is not a part of the end times because we are not quite there yet. We are on the cusp. But it is definitely a precursor. A sign of things to come. Wake up if you haven't already punched your ticket for eternal life. True Christians get to look forward to the rapture which is going to come before the real horror hits the fan. Get them Evolution and big bang and baby killing Democrat lies out of your head. Cuz after you die it'll be way too late.

-BREAKING NEWS!!! BREAKING NEWS!!! Human biological scientists, along with gender studies and a number of college psychology professors, have determined finally that there absolutely are more than two genders. And that is now a scientific fact (just like evolution). They have determined that of course there is the male and the female gender which we are all aware of. But now there is a third gender, and includes all of those geniuses who think there are more than just two genders, and that third one is called the shit-for-brains gender. And do not make the mistake of calling these retards either male or female when you address them. You must address them either as Mr. or Mrs. shit-for-brains. If you don't you could be either arrested for pronoun misappropriation, or because you're not a Democrat voting jackass. Stay tuned to this channel for further updates.

-Kingdom theology is a pile of buffalo dung. No one's going to build any Kingdom. This world is going to plunge further and further into a hell on earth ruled by the baby-killers (same as the Jew killing Nazis 80 years ago in Germany) and the clitoral mutilating, head-chopping, homosexual killers, with their one world government where Satan and his antichrist will rule for a 7 year tribulation period at the end of which the Lord returns in the clouds of heaven and cleanses the earth of their stench.

-Hi! I am Mark Suckdickerberg. I am a left-wing indoctrinated, mentally poisoned, baby killing, racist against the black race, and black baby exterminating useful idiot of Satan and stuck on his Democrat Plantation.

And I am doing everything I can to mentally poison as many more people as possible to keep them stuck on thefarm, so my father Satan can suck all of those poor unfortunate lie-filled geniuses down to hell with him forever and ever, burning alive and screaming at the top of their lungs. Thank you Democrats for your help in this evil satanic endeavor. I can't tell you how much I appreciate it.

-"I have seen a wicked and ruthless man flourishing like a green tree in its native soil, but he soon passed away and was no more. Though I looked for him, he could not be found." ....In case anybody is wondering where Elijah Cummings, Teddy Kennedy, John Lewis and Hillary Killery (I know, but she's as good as dead.) and their kind are right now. If you go to hell when you die, and I pray that you don't, but if you do say hello to them for me. God will hold the baby mutilators that sacrifice his babies to Satan, he will hold them in absolute derision. "The Lord laughs at the wicked, those who mutilate his babies to death and vote for the same, for he knows their day is coming..." Better wake up!

-Here's a neat story. There is this massively successful Church in Glen Burnie Maryland (called Lighthouse). And whereas I don't have anything negative to say about it. I have approached numerous of the workers there, like they have a desk where they answer questions, and they have greeters all over the place. And I have asked a number of them if it is possible to be a true Christian on their way to an eternity in heaven, and yet at the same time vote for the baby killing Democrats. And you would be surprised at some of the responses I've gotten from this great church of Jesus. I would say half are like, "yeah, exactly, I know what you mean". But the other half are like they look at me quizzically like they have never even broached that thought before in their brain. Maybe because they vote Democrat. IDK. But one nice black gentleman, a greeter at the door where you walk into the sanctuary, replied to my question by saying "what does faith have to do with politics". **What DPES FAITH HAVE TO DO WITH POLITICS??#$!!?** Are you freaking kidding me??! Faith has absolutely everything to do with politics. Faith has absolutely everything to do with everything!! Like if I have faith in Jesus Christ to take me to heaven when I die I am not going to vote for baby mutilators. I am not going to vote for murderers because if I do that makes me one too. And there won't be any murderers in heaven. Unfortunately I didn't have a chance to tell the gentleman that cuz all he did was try to shut me up by constantly repeating "what does faith have to do with politics." Unbelievable. "Hear no truth. See no truth. Speak no truth."

Obviously this is a church, whereas they address so much about the love of Jesus, they haven't spent enough time talking about the truth of Jesus. Because the love of Jesus without the truth of Jesus produces people like that gentleman standing in greeting at the door to the sanctuary. Obviously a Democrat voter, and someone who is obviously on their way to an eternity in hell. Maybe somebody should actually take the time, and have the balls, to witness to that gentleman. Maybe from the pulpit?? I mean you can get really successful at playing Jesus. But I thought we were supposed to be the salt of the earth. And that gentleman has a large wound that is crying for some salt. Yet he is in a great Christian Church and nobody is offering him any salt. Why, Lighthouse? Why?

-Dec. 2nd, 2023: Trump: "We Know The Results of the 2020 Election!" Yes, and so did all of us with a brain, right from the day after the election in November 2020. It was clearly stolen. He clearly won. That's why I completely stopped watching Fox News then. It was nauseating listening to them going along with the Demon RATS theft. But why did the Lord allow that theft to occur, and for Trump to effectively lose?? In my opinion it was because Trump did NOT defund Planned Parenthood. He listened to all of his advisors that said OH NO!!!he couldn't do it. He couldn't shut the government down because the Nazi lying Democrat media would blame it on him and the Republicans. But he should have, and then the Almighty God of all creation would have taken care of the Demonrats and their media. But he didn't. Big mistake. And we all have suffered the horrible consequences ever since...

-Here's something else that the evolutionary indoctrinated, intellectual idiots that own our educational system and mock God and mock the Bible, need to have put in front of their faces. The evidence for that global worldwide flood that happened 4500 years ago according to that same book, is absolutely overwhelming. The Grand Canyon could only be formed by a massive run off of water the size of one of our Great Lakes. The White Cliffs of Dover, the massive sand dunes at Kitty Hawk, seashells on the tops of the mountains all around the world, and the existence all across the Earth of sedimentary layers thousands of feet deep that cover thousands of square miles could only have been formed by that massive, world-wide flood. (You're welcome.) (I'll send you my bill in the morning.)

-Do you have an itch on the top of your head right now? Yes? No? .... Well if you do, and even if you don't, go ahead and scratch it for a second.

And now understand that the same level of difficulty and effort it took for you to scratch your head is the exact same level of effort and difficulty it took for God to send the 10 plagues on the nation of Egypt back in ancient times. And that it is the same level of effort and difficulty it took for Jesus to raise the dead or to walk on water or to feed over 10,000 men women and children with just five loaves of bread and two fishes. In fact, it is the same level of effort and difficulty that it took for Jesus to speak all the worlds into existence, and all the living into being on this planet, about 6,000 years ago. ....And now you know the rest of the story. BTW. Evolution is for idiots. And Atheism is for idiots. With all due respect of course. Sorry if anyone got offended. But if you did get offended then you need to come back when you get a little more actual edgumakayshun. The truth can be very offensive, to people whose minds are poisoned with lies that they were taught all their life.

-Pantheism is false. God is not the world. God is a spirit that projects the world into our brains. Because the world is just an image in the mind of God. But the image, as wonderful and beautiful and glorious as it is, is not God. And God is not limited just to light even though He IS light. God permeates every square centimeter of this universe with His presence and His being. So He is certainly not limited to just light. But at the same token, God IS light. ...(Get something to drink and then reread it. It might make a little more sense.) ☺

-To all of my evolutionary indoctrinated friends all over FB, I can assure you there is as much evidence for the theory of evolution and also for the theory of the Big Bang, there is as much evidence for those two theories as there is for the theory that big bird built the pyramids. Now I know you think that's far-fetched but it's absolutely true. You just have not read all of the information that I have. I took all the information from 20 to 30 books that totally destroy Evolution and the big bang and are written by some of the world's top scientists. Unfortunately they are all blackballed from the mainstream Democrat media, educational system and entertainment industry that has a vested interest in perpetrating the LIE of evolution because it is the quick sand foundation that props up their baby killing ideology from hell. In fact I wrote a book titled The Theory of Evolution is a Joke. It is a hundred and twenty pages and it has 250 quotes from some of the world's leading scientists. The information and the facts and the evidence is absolutely overwhelming. And once you have that information inside of your brain you no longer are mentally poisoned and mentally deranged with all the lies that we were all taught. You then see those two piles of stinky scientifically impossible horseshit for the absolute joke and the absolute drivel that they both are. And that they have absolutely nothing whatsoever to do with science or reality. But they have a lot to do with nonsense. So there you go. Good luck and God bless.

-And the account of the creation and origin of the universe and life on Earth that can be found in Genesis the first book of the Bible is absolutely scientifically and historically accurate, and is based on all of the evidence. We were lied to by Liars who were lied to themselves. Wake up! You've been lied to. We were all lied to.

-Evolution actually is for idiots. And of course we all were idiots, weren't we? We all were taught it and we all believed it. But it really is for idiots, and here's why.... Which Evolved first? The 60,000 miles of blood vessels in your body? Or the blood that goes inside of those vessels? Cuz one of those without the other is a waste of time because the organism dies. They both have to evolve together at exactly the same moment in time in exactly the same body. No hundreds of millions of years of small incremental steps for that skippy. So your blood system and your blood vessels they are indistinguishable from a miracle aren't they? But that's not evolution. A miracle like that has to be created doesn't it? Of course it does. The same with your eye and your eyelids. Which came first. Why would an eyelid evolve without an eye? That's an idiotic waste of time. But if your eye evolved without the eyelid, as well as the tear ducts evolving simultaneously, the eye dries up and dies. Again indistinguishable from a miracle isn't it? Which has nothing to do with those slow incremental steps over hundreds of millions of years. Same with your musculature throughout your body without all of your perfect exactly positioned bones. Which evolved first geniuses? They have to both evolve together at exactly the same time, and place, in exactly the same body. Again, no incremental steps over hundreds of millions of years. Etc. Etc. And on and on.... ...I just destroyed Evolution, didn't I? Of course I did. Anybody can. SCIENCE CAN. AND DOES, You see, evolution really is a lie. And it was foisted upon all of us, wasn't it? The only thing that remains is whether you want to remain an idiot, or you want to free yourself from that idiotic lie. Good luck and God bless. Oh, and when you do free yourself, after completing the 12 steps, and you attain logic and common sense, please let us know here at Evolutionics Anonymous. (EA). We sure appreciate it.

-Isn't it hilarious when you run into people on Fascist-Censor-Book who have been lied to all of their life in the Democrat evolutionary indoctrination brainwashing educational system, just like we all were lied to, but when they have the truth put in front of their face on here, it's hilarious how they spaz into a hissy conniption fit for the ages. It's so amusing. They are called willfully ignorant. And they were taught by intellectual idiots. And for God's sake, we're just posting the truth. But they've got to protect those stupid lies in their head. At all costs.

-There are literally thousands upon thousands upon countless thousands of actual facts and evidence that totally annihilate and destroy both the theory of evolution and the theory of the Big Bang. That's what science says. Real science. Not the faux science that actually gives those two ridiculous idiotic theories a total pass. Because they're both nothing but religions.

-The brain dead Evolution believing God haters and God mockers have come up with what they think is a really brilliant put down to those who are not as stupid and willfully ignorant as them, and that would be their referring to the real Creator of the universe and of all life on earth, as a "Sky Daddy." As opposed to their idiotic, retarded and non-existent god of random accidental chance. That's their "Earth Mommy." And in that reference to a sky daddy they get to dis God, and fathers, at the same time. Two things that the baby mutilating Democrats hate the most. Of course they learned that term from the fanatic evolutionary Neanderthals that indoctrinated them in high school and college. They are so smart. …The level of knuckle-dragging ignorance it takes to believe that life just made itself is actually quite remarkable, thank you Richard Dawkins and Christopher Hitchens. ...."Sky Daddy." HahahahaHAH!! ...Horses asses...

-OK. A little information for my poor unfortunate evolutionary indoctrinated friends, who still believe that total horseshit. Question... When did dinosaurs go extinct?? How many millions of years ago was that?? Wrong again. Dinosaurs and dragons are one and the same. Ancient writings from throughout Europe and England and Africa and the Far East have numerous accounts of dinosaurs that are called dragons that lived alongside man in our own historical times. Look it up. In fact, here's something out of an ancient book that talks about dinosaurs walking among men. Not millions and millions of years ago like the evolutionary indoctrinated have taught us. And I quote, "Look at the behemoth, which I made along with you man, which feeds on grass like an ox. What strength he has in his loins, what power in the muscles of his belly. HIS TAIL SWAYS LIKE A CEDAR." [Emphasis mine] Now, what could that possibly be? An elephant? Wrong. An elephant's tail is like a rope. The only thing that can be is a brontosaurus. I will continue from that ancient book, "When the river rages, he is not alarmed. He is secure, though the Jordan should surge against his mouth. Can anyone capture him by the eyes, or trap him and pierce his nose?" Of course they can't. How could ancient man ever trap a brontosaurus? And what other animal could walk through a raging River without a care in the world? Dinosaurs lived at the same time as man 3, 4 and 5 thousand years ago. And now you know the rest of the story. I'll send you my bill in the morning.

-Did your paperclip make itself? Of course not! Only an idiot would think that. It had to be made of course. And could your paperclip ever evolve over hundreds of millions of years into a safety pin? Of course not! Only a total moron would think that. Well guess what my friend. You just destroyed evolution. Life is a trillion times more complicated than a paperclip.

-This is for my evolutionary and Big bang indoctrinated friends... Question, what kind of a solar system exactly would the universe create accidentally by random chance if there was no such thing as creation? Well that's easy to answer. Just get yourself a little box, go outside in your yard and fill it up with rocks and dirt. Then take it over onto your driveway and fling it out. There you go. That's the kind of solar system you would have gotten if the Big Bang Theory was real and an explosion created the universe. Of course it is a scientific fact according to the laws of Science and all the facts and evidence that the theory of evolution and its companion theory of the Big Bang are both Total piles of horseshit. They have nothing to do with science. Evolution really is a joke and it really is for idiots. And so is the big explosion. Especially now with all that we have learned and all that we know scientifically. Neither theory could create a damn thing. They are both the laughingstock of the universe. Wake up dummies you've been lied to.

-Darwin Sunday??!!$#!%$#?! What can be said about Darwin Sunday? Well, here goes.... Darwin Sunday is a once yearly pile of Jesus hating vomit that was created by a number of brain dead fake Christian churches that are a stench upon this once great land. On that pathetic Sunday a bunch of brain-dead Pastors along with their brain dead congregation apologize to Darwin, I guess because real Christians weren't stupid enough to believe his scientifically-impossible theory. That these brain-dead fake Christian suckers actually believe. Of course they believe the lies of Satan while at the same time they call themselves Christians. But they obviously hate the truth of Jesus Christ. They are Satan's suckers. (They are akin to that pathetic "Christian" satanic church somewhere out in the Midwest where Tiller the baby killer and his wife were ushers and elders). And that's the truth about Darwin Sunday. And I say this, with all due respect of course.

-We live in a spiritual and not a material universe. There is nothing material here. Even the building block of all matter the atom is 99.999% empty space. And the electrons and neutrons and protons that comprise the atom are not hard little marbles, but themselves are nothing more than energy units whatever that is. The entire universe is nothing more than an image in the mind of God. And nothing less. If you build a road out of asphalt it is made of asphalt. If you build a house out of bricks it is made of bricks. God made the universe out of nothing, so it is made of nothing. Emptiness here. Emptiness there. There is no such thing as materialism.

-What are bugs? You know, flies, bees, all the little flying things, ants, spiders and all the little crawling things. They are nanobots. Tiny little robots. The greatest robot maker of all preprogrammed them and then spread them across the world 6000 years ago and they've been self-replicating and living and feeding and flying & crawling and dying ever since. Mankind with all his lofty intelligence has spent trillions of hours trying to make robots and they have come along very impressively. But even if you gave them another thousand years they would not be able to replicate or duplicate what Jesus the Lord imagined and envisioned and then created and placed on this Earth 6000 years ago. Our educational system might just want to give Him some credit. Right after they pry their heads out of their asses. Their mentally poisoned, brainwashed, evolutionary indoctrinated heads, out of their non-evolved asses.

-Someone asked why I don't explain how the universe and life on Earth actually *did* come into being, rather than just destroying evolution and the Big Bang and stating how it did not. Good question. Answer..... The problem is nobody who actually believes in evolution and the big bang that we were all taught, is in any position whatsoever to listen to the truth historically and scientifically about how all this did come about, until those idiotic ridiculous and scientifically impossible lies are exposed and therefore destroyed. So of course that is my initial focus. But yours is a very good question so here goes. ...6000 years ago the Lord, Jesus, God, who is a spirit that is not contained within space or time, and existed before He created the universe as well as space and time, He spoke into creation the entire universe. But He started with a ball of water 2 light years across. Now that's not very much, two light-years when you compare it to the billions of light-years that the Universe has been stretched out to now. But understand that the universe is 99.999% empty space. And a ball of water two light years across is large enough to contain all of the matter of the known universe. After he created that ball of water he stretched it out as he created all the stars and all of the galaxies. Hence the Doppler effect and the reason why people have mistakenly said that the Universe has to be 15 billion years old cuz that's how far away light has had to travel in order for us to see it. But that's absolutely not true. When the Lord stretched out that ball of water while creating the stars and the galaxies the light was also stretched out. Hence the Doppler effect. And if you read the beginning of the book of Genesis in what has very credibly been called God's word because it says over three thousand times it was written by Him and it's got thousands of miracles and thousands of fulfilled prophecies in addition to many other evidences to back up that claim, but you will find that it talks about how the spirit of God hovered over the face of the deep, that initial ball of water two light years across. And isn't it interesting that water is life and that he would start off with the creation of it. And there have been scientists that using that beginning have predicted the magnetic fields of our planets accurately before they were discovered. Interesting. And then over the next few days Jesus spoke all the living into being on this planet. And that is the accurate scientific and historical record of how it all came about. Good luck and God bless.

-Satan and his demons as well as his followers here on earth, they are where they want to be. They have freed themselves from the love of Jesus. And that way they are not beholden to Him. They have become their own gods. Which unfortunately is no god at all.

(to be continued…)